CONSTRUCTION CONTRACTORS' SURVIVAL GUIDE

CONSTRUCTION CONTRACTORS' SURVIVAL GUIDE

THOMAS C. SCHLEIFER

WILEY

A WILEY-INTERSCIENCE PUBLICATION

JOHN WILEY & SONS

New York • Chichester • Brisbane • Toronto • Singapore

The pronoun "he" is used to refer to contractors of both sexes. The author and publisher chose this convention for stylistic purposes, to avoid the more cumbersome "he or she" or "he/she" constructions that prove awkward.

Library of Congress Cataloging in Publication Data:
Schleifer, Thomas C.
 Construction contractors' survival guide / Thomas C. Schleifer.
  p. cm. — (Wiley series of practical construction guides, ISSN 0271-6011)

 "A Wiley-Interscience publication."
 Includes bibliographical references.
 ISBN 0-471-51324-5
 1. Construction industry—Management. I. Title. II. Series.

HD9715.A2S365 1990
692'.8—dc20   89-38321
              CIP

10 9 8

*To Sophann,
my wife and inspiration*

SERIES PREFACE

Congratulations! You've just bought a profit-making tool that is inexpensive and requires no maintenance, no overhead, and no amortization. Actually, it will increase in value for you each time you use this volume in the Wiley Series of Practical Construction Guides. This book should contribute toward getting your project done under budget, ahead of schedule, and out of court.

For nearly quarter of a century, over 50 books on various aspects of construction and contracting have appeared in this series. If one is still valid, it is "up-dated" to stay on the cutting edge. If it ceases to serve, it goes out of print. Thus you get the most advanced construction practice and technology information available from experts who use it on the job.

The Associated General Contractors of America (AGC) statistician advises that the construction industry now represents close to 10% of the gross national product (GNP), some 410 billion dollars worth per year. Therefore, simple, off-the-shelf books won't work. The construction industry is unique in that it is the only one where the factory goes out to the buyer at the point of sale. The constructor takes more than the normal risk in operating a needed service business.

Until the advent of the series, various single books (many by professors), magazine articles, and vendors' literature constituted the total source of information for builders. To fill this need, this series has provided solid usable information and data for and by working constructors. This has increased the contractors' earning capacity while giving the owner a better product. Profit is not a dirty word. The Wiley Series of Practical Construction Guides is dedicated to that cause.

M. D. MORRIS, P.E.

Ithaca, New York
November 1989

PREFACE

Your management decisions alone determine whether you will succeed or fail in the construction business.

Many contractors believe that they lose money or fail because of weather conditions, labor problems, inflation, unexpected rises in interest rates, the high costs of equipment, a tightening or shrinking of the market, or simple bad luck. Actually, none of these has ever been the primary reason for contractor failure. They may have contributed to failure once a bad management decision was made. But they were not the basic cause of the failure. To put this another way, we, as contractors, often attribute our failure to make profits or even our failure to survive to conditions over which we have no control. Actually, that is not the case.

If you think that because you are surviving or even thriving in the industry you can drop your guard, you may be wrong. Not only growth and expansion, but even ongoing operations mean changing. And change can make or break you.

Past success is not an indicator of future success, again, for management reasons. When a company expands in size, takes on larger projects, or goes after projects of different kinds or in different territories, it requires management decisions to reduce the risks inherent in such change. For example, you may be doing fairly well, or even very well. But with the stress to an organization of growth or change, a weak component can become a fatal component. And change itself has to be managed to minimize risk.

This book identifies the ten elements in contractor failure and shows you how to avoid them.

Contractors don't talk to each other. Oh sure, we meet, converse, socialize, make jokes, and tell stories about ourselves, our friends in the trade, our competitors, but we don't tell each other about how we run our businesses. We don't talk about the mistakes that cost us profits, deadlines, or, at worst, our entire operation. The truth of the matter is that there is a code among contractors. It's an unspoken code, but virtually everyone knows it or learns it fast enough. The way the code works is that you learn the basics of the industry, you acquire the essential information and knowledge of the industry on your own through personal experience. Once you've gained that information, it's yours. You've paid for it. You've earned it. It's your personal property. Let others earn it and learn it the same way. Because of this attitude, contractors continue to invent the wheel every day.

By pointing out this code I don't mean to disparage it. There are some very good reasons for its existence. If you include those people involved in ancillary jobs or in the manufacturing and transporting of building materials, there are more people involved in construction in this country than in any other industry. It's a $400-billion industry with more than 1.4 million individual businesses. The turnover rate as companies go out of business and start-ups replace them is phenomenal.

Construction is doubtless the most highly competitive and high-risk industry in the United States. Consequently, if a contractor can survive a mistake, correct it, and learn from it, he knows something his competitor might not know. This information gives the contractor a competitive edge. He's not likely to give that edge away.

So there are good reasons for the code. Unfortunately, the code has some very negative spinoffs. Without cross-fertilization, a sharing of essential information, and the collection of a body of knowledge available to the industry as a whole, there is a significant time lag in improvements and modernization within the industry, in the way we operate our businesses, and in the way we manage our enterprises.

To put this in perspective, think of the other industries that compare to construction in terms of their contribution to the gross national product—the automobile, steel, oil, and aircraft industries. All of them have extensive training programs that train their personnel from entry to top-management levels. And these training programs are ongoing and under continual review and revision. These industries have also developed a system of checks and balances on their decisions and strategies; they have boards of directors to ensure accountability and monitor managerial decisions and techniques. In contrast, nine times out of ten the start-up contractor has learned how to run a construction business by watching, by working various jobs, and by working with someone who has been successful in the business. Few are actually taught how to run a business. There is a hard grain of truth in the old construction story that all you need to be a start-up contractor is a pickup truck, a hard hat, a box of tools, a cast-iron stomach, a forgiving wife, and a bad temper. Many have started with less.

There is no *one way* to run a construction business. But there are only a *few* ways to run one successfully.

Typically, people in the construction business assume that there is no method to be taught because there are as many ways to run a construction business as there are construction businesses. The tragedy to the industry is that this is a false assumption. True, each business develops its own style. But there are only a very few ways to run a successful construction business.

There are methods that can be learned and must be employed to control the risks in this extremely high-risk business, to allow the contractor to take informed risks, and to allow the contractor to learn from the mistakes of others as well as his own. Unfortunately, major business strategy mistakes in the construction business are usually fatal to the contractor, and until now, no one has collected sufficient data on the subject to give us the hard facts.

This book is the result of over 30 years of experience in the construction business, 11 as a contractor and 10 as a consultant to hundreds of distressed or failing construction companies. In the process, I have been able to collect enough information on the causes of contractor failure to categorize them and use them as a learning tool. From these examples I have been able to isolate the ten elements of contractor failure. One or more of these ten elements has appeared in every business failure I have studied.

After an overview of the industry and a general presentation of the ten elements of contractor failure, subsequent chapters will take the ten common elements one by one. The chapters will define the element of failure clearly, give real examples, and discuss ways to minimize risks involved. A recognition of these elements can be a tremendous asset. After the ten elements of failure have been discussed together and separately, the last chapter will tie it all together by showing you how to take these elements of failure, these don'ts, and use them to develop a positive and competent management attitude and strategy.

We like to believe that sheer energy, drive, ambition, know-how, and guts will get us through in the construction business. And it probably will for a while. But there comes a time when that energy and drive have to be organized, given direction, planned, held to objectives. If you don't have a proper organization, you'll be in trouble.

A contracting business has only three primary functions: getting the work, doing the work, and accounting for the work, or put in other terms, marketing, operation, and administration. The first thing a contractor needs to understand about the business is that these three functions exist and that they are separate and distinct from each other. Sure, they overlap and intersect. But they can only be dealt with properly if they are treated separately. Once your business and technical activities are broken down into these functions, and once your time and energies are budgeted to treat them separately and properly, you will be organized. Once you are organized, you can manage the organization.

One person must have direct personal responsibility for each of these three functions. This can mean three different people or two or one. It is not unusual for a small company to have one person handling the estimating and bidding and the accounting and billing while the other handles the construction operations. Neither is it unusual for one person to handle all three functions. But as your business grows, you have to remember to keep the functions distinct.

The important fact is that the personal responsibility must be clearly recognized. One role may appear more important and significant to the operation of the company. But that is mere appearance. If you neglect any of these functions, you are courting failure. They are essential.

Before going further, I want to define my use of the term "contractor." The contractor is easy to identify in small and some medium-sized companies. He is the owner of the construction company, the person who contracts to do the work. In other medium-sized and larger firms, the designation of specific individuals among top management as contractors is more difficult. In my view, the people who accept responsibility for one or more of the three primary functional areas of management described above are contractors, whether or not they own a piece of the company. The key word in my definition of a contractor is "accept." The individual must believe he is ultimately responsible for the success of his functional area of the business. If he accepts that responsibility personally—not as a functionary or an executive, but as a principal in the organization—then I define him as a contractor. By contrast, if an individual thinks of himself and refers to himself as the chief estimator of the chief financial official, he probably isn't a contractor.

I begin with such basic organizational principles because many contractors believe that quality construction and a few breaks should guarantee business success. They don't. I believe that identifying the elements of past failures will provide you with the means to achieve future success the way generals review past battles and the way medical researchers study illnesses to find ways of prevention and cure. If any good at all can be said to come out of the tremendous number of contractor failures over the past years, it would be to pass along the lessons learned from their mistakes. This book is dedicated to the fine builders who didn't make it, and the information is offered in the name of those good men and women.

Thomas C. Schleifer

Mendham, N.J.
April 1989

CONTENTS

CONSTRUCTION CONTRACTORS' SURVIVAL GUIDE

CHAPTER 1

Managing with Confidence

1.1 LESSONS LEARNED

Failed contractors worked in the same environment in which others of similar size and experience succeeded; outside stimuli did not cause their failure. Some of the more obvious industry problems like weather, labor problems, inflation, and even a fluctuating marketplace aren't enough to put a construction company out of business. These contractors had freedom of choice to manage as they saw fit, and they failed by virtue of their choices. Most importantly, there are enough cases for us to see distinct patterns in decisions and other management actions that occur so often they can be identified as common causes of contractor failure. We learn from the study of these companies some extremely important lessons about what not to do as well as a lot about what we should do. If we understand in advance the risks to our success and if we know when they are likely to occur, we know enough to avoid or prepare for them.

1.2 NO GUARANTEES

This study reveals a very disheartening fact: success in the construction business, even for very long periods, doesn't guarantee continuing success. In fact, the study indicates clearly that every change in a successful organization, particularly growth, creates a period of risk in spite of all previous successes. Even a carefully thought-out strategic plan cannot eliminate risk but can only highlight it and prepare for it.

1.3 MANAGING AREAS OF RISK

The good news is that while the risks are many, they *can* be dealt with and managed. This study has identified the ten areas of greatest risk to contractors.

Any one of these components handled badly can create problems for a contractor but should not put him out of business unless he is already in serious financial difficulty. All changes in business cost money and perhaps even shake up an organization, but dealing with any one of these areas of risk at a time should not be life threatening to a healthy business even if the change isn't handled well.

Most of the failures I have observed stemmed from a contractor's attempting to deal with two or more of these components at one time. Even with good management skills it is extremely difficult to steer a business around several obstacles at once, and each of these areas presents serious obstacles to success. If a contractor, while working on one project, takes a much larger project at a distant site, the degree of difficulty or risk may be easy to see, but he will also have a serious risk if he has inadequate equipment cost control and poor billing procedures when work starts to slow down, and this compounded risk may not be so evident. Adding another component increases his risk geometrically.

Every contractor will face each of these areas as his business grows and outgrows his organization. Knowing you will face these aspects of growth, you can do your best to have to face only one at a time. If you must manage several changes or improvements together, you will need to delegate a great deal of authority, and doing this will test your managerial maturity.

Consider these recommendations, based solidly on contractors' experience. What they learned can reduce your risk.

Increase in Project Size (Chapter 3)

Increase project size gradually.

Doing a project twice the size of your current top size is difficult.

The risk is proportional to the change.

Take only one larger project at a time.

Finish the first larger project and evaluate before taking the next one.

Don't leapfrog and keep doubling the size of projects.

Unfamiliarity with Distant Geographic Areas (Chapter 4)

Stake out your best area.

Expand slowly away from it.

Test a new area with a small job.

Heading out in only one direction may be helpful.

Plan carefully for expensive regional offices.
Know when to withdraw.

Replacing Key Personnel (Chapter 6)

Decide who is key.
React quickly to replacement need.
Carefully test size replacements.
Carefully test replacements.
Pare down as a safety valve.
Watch for burnout.

Managerial Maturity (Chapter 7)

Management is always tested during growth.
Understand management needs of a growing company.
Learn to recognize signs of inadequate management.
Lack of managerial maturity can happen to anyone.
Get help in time.
Delegate.

Accounting Systems (Chapter 8)

Accept responsibility.
The reports should make sense to you.
Expand or improve for growth stages.
Timing is everything.
Don't let them fool you.
Don't neglect your gut response.

Evaluating Contract Profitability (Chapter 9)

It's all that counts.
Each job stands on its own.
Select the system yourself.
Discipline the information flow.
Evaluate monthly and accurately.
Use evaluation as a tool.

Equipment Cost Control (Chapter 10)

Even paid-for, nonworking equipment costs money.
More is not always better.

Bigger is not always better.
Idle time is a job expense, not overhead.
Account for replacement.

Billing Procedures (Chapter 11)

Contractors cause half the problem.
Invoice on time.
Be entitled, not embarrassed.
Ask first, then demand.
You're not a bank.

Transition to Computerized Accounting (Chapter 12)

The data come from you.
Computers are not failsafe.
Change requires planning.
Test electronic data before relying on it.
The reports should make sense to you.
Have a backup system.

Other Industry Concerns (Chapter 13)

Be profit driven, not volume driven.
Plan your cash flow.
Develop some flexible overhead.
Make employee benefits realistic.
Tie bonuses to performance.

1.4 RECOGNIZING SIGNS OF POTENTIAL TROUBLE

Attending to the following indicators of a possible underlying problem can also help keep a company on track:

Unexplained tight cash flow
Decline in profit margin
Disproportional increase in overhead
Increase in the turnover of personnel
Increase in claims activity

Late accounting information

Changes in accounting information reported

Unexpected borrowing

Increase of internal disputes

Decrease in the quality of work

Too many excuses

Departures in the accounting staff

Inadequate time to do anything well

The presence of one or more of these conditions can have very logical and legitimate explanations and indicate nothing more than a bump in the road. They are simply signs of more serious organizational problems that went unaddressed by contractors who either didn't recognize them or didn't have the management skills to deal with them. Any of these conditions is worth a hard look. Resistance from staff to necessary fact finding is a sure sign that something is wrong. The problem may, of course, be in the department being questioned. But when middle managers or other top managers want to keep their areas of responsibility to themselves and resent even a nonjudgmental inquiry, a problem exists. A well-run business fosters an open attitude that encourages everyone in the organization to be well informed and to pull in the same direction. Inquiries by top management should not be viewed with suspicion.

1.5 LAYERS OF MANAGEMENT

Another problem is that top management of a growing organization can become separated from the mainstream of the organization to such a degree that the indicators of trouble become invisible to them. This can and does happen in organizations of all sizes for various reasons. Sometimes the contractor takes so much of the work load upon himself that he has very little time to manage well and no time to plan or review progress or performance. He never steps back to see how the company is doing, to get the big picture. This happens in small and large organizations with the same frequency. Other contractors are unaware of trouble signs because their middle managers don't report them, and top managers of large companies cannot observe all details at all times. And still others are just not observant. They're great at putting construction in place but not skilled at overviewing the business and being alert to subtle changes that can affect it.

1.6 OWNER VERSUS TOP MANAGEMENT

The fact that in the construction business a company's owner(s) and its top manager(s) are usually one and the same creates a unique management problem.

The owner of a construction company is the ultimate party at risk and, as such, should set the goals and objectives of the company. Management from the president on down is responsible for carrying out those goals and to conduct the day-to-day operation of the business in compliance with them. It is the owner's role to see to it that the goals and objectives of the company are accomplished by management.

In most construction companies, no one is playing the important role of the owner. The owner-manager can wear both hats but not at the same time even though he may believe he is doing just that. A contractor has a choice—to work alone or get help. If you decide to work alone, you should set aside certain times, perhaps three or four times a year, to take off your manager's hat and put on that of the owner. This shouldn't be done during the day-to-day activity or with interruptions to do management work. If you select your office as the place to do the owner's job, evenings or weekends will work best unless you are able to completely avoid interruptions. And even if you can be assured of uninterrupted time, you may not be able to force your mind away from the day's pressing management tasks. If you have partners or minority shareholders, they should be included to the extent that they understand an owner's role as distinct from their management responsibilities.

1.7 DISCIPLINING PERFORMANCE

The task before you is to review corporate goals and to critically scrutinize the company's performance—*your* performance—toward accomplishing them. You're checking on yourself, but this can be meaningful if you work at it. You need to forget about how hard you have worked to accomplish your objectives and about everything that has stood in your way and objectively evaluate your progress. Ask yourself whether the company is moving in the direction you had decided it should the last time you wore your owner's hat, and if it isn't, ask for reasons. We all need to be accountable to someone, and if you choose to be accountable to yourself, you should go through this exercise regularly and make it work. Don't fall into the trap of believing that you are doing this every day or on a regular basis. This process must be objective, and that requires discipline and mental distance from daily activities.

A critical performance analysis and the disciplining of the push toward goals and objectives are missing ingredients in the formula for success in almost every closely held corporation. This lack of management accountability is one reason for the high number of business fatalities in the construction industry. It's not only lonely at the top, it also doesn't make good business sense to run an entire organization without some independent outside verification of your strategies.

1.8 BOARDS OF DIRECTORS

I very strongly recommend to contractors of every size that they have an active board of directors. I don't mean a board in name only or one that meets for dinner occasionally, and I certainly don't mean one made up entirely of family members or insiders. A board of directors represents the owner's or stockholders' interest and can be particularly useful where the owners and management are the same. What seems to make sense in the day-to-day management of the business may not in fact be good for the company in the long run, and your board can help you keep the company on track.

When my former firm had only 12 employees, it also had a three-person board of directors with one outside member. I wanted people who were successful in business but not necessarily in my business or even in my industry. Later the board consisted of four outsiders and three insiders, and I owe a great deal of the success of that business to the outside members. They asked the hard questions, set some of the tough goals, called me on my mistakes, and kept me accountable. During most of that time I was the only stockholder.

1.9 ACCOUNTABILITY

Board members or advisors from outside the company or industry provide the missing ingredient in the closely held corporation—*accountability*. They protect the company's assets by asking hard questions. Because they aren't caught up in the day-to-day activity of the business, they are able to see the whole picture.

The formation of a board of directors to help control risk makes good sense and is no threat to a contractor's rights as an owner or to his independence. In the current insurance market you'll find directors' liability expensive or unavailable, so you may want to form a board of advisors. You can get legal advise on structuring the board to limit or eliminate exposure to the outside members. If you can't deal with the idea of more outside members than insiders at first, then start with a small number, say, five, and have only two outsiders. An odd number of board members is customary to prevent ties, but in actuality there are more discussions than votes in a closely held company, and an even number works as well. What is most important is that the outsiders be truly outsiders, not the corporate accountant, attorney, or banker and not necessarily close friends—unless you know them to be objective enough to participate and speak their minds regardless of the friendship. The only prerequisite is that they be successful business people. They need not be knowledgeable about your industry or your business. In fact, it is preferable to have each outside member from an industry other than yours and different from the other members. If you do put outside members on a working board, it will be ineffective for family members or nonactive board members to continue to sit unless they are qualified business people and can interact as peers.

1.10 SELECTING THE MEMBERS

You can select the right people from the business community in which you live or work. The size of your organization should be a guide in selecting candidates with helpful experience. Most business people are complimented by being asked even if they haven't the time. Candidates need to know how much time you're asking for and that you intend to have a real working, voting board so that their input will be important. Of course, you will need to interview them if you don't already know them because you should be compatible and comfortable with them.

Professional board members are paid and well worth the expense. For small companies $300–$500 a meeting may be enough; large companies may spend several thousand per year. Some companies pay by the meeting and some by the year plus each meeting. Many companies do not pay additional compensation to inside board members, although some do.

Your board should meet as often as you feel you need to; meeting less than twice a year means the members will lose touch, and meeting more often than once a month means they will become more like management than advisors. Asking each member to sit on the board for a specific term allows you flexibility in membership without embarrassment. The bylaws of your corporation will set out the minimum duties of a board of directors, but you should emphasize to the members that their main function is to set corporate goals and to evaluate and discipline the achievement of those goals.

An active board of directors with outside members not only will protect your assets and make your job easier, but also can be a great source of strength when you're riding out tough times. It's one of few business decisions involving no risk.

1.11 CREDIT CONCERNS

Construction contractors generally rely on a steady source of credit for their continued success. Secured equipment loans and unsecured working capital lines of credit as well as surety capacity or credit to secure bid, payment, and performance bonds are the life blood of many construction enterprises particularly those that are growing. That is to say that without continued bank and bonding credit many construction companies would cease to exist very quickly. It is as necessary an ingredient for success as management ability, manpower, and brick and mortar. Herein lies a problem for the contractor in this high-risk business. Continued credit relies on, among other things, favorable performance as demonstrated or depicted in the annual financial statements of the company. Put another way, credit is fairly easier to get when things are going well but can get difficult and even dry up during tough times. One bad year for some contractors may not cause too many problems in this area, but two or more bad years in a row can restrict or even dry up the credit necessary to continue in business. A risk for the contractor then is to have internal or external events affect his business such that he cannot reverse losses within one or two fiscal years, three at the most, or before it adversely affects the credit so necessary to the continuance of his business.

Taken one at a time, the common elements of contractor failures described in the next chapter may seem too subtle to cause a business to go under. But when taken in the context that they were enough to cause many businesses to lose money for two plus years, the risks become more apparent.

In this credit-centered industry the contractor faces the risk of losing money and the risk of losing or restricting his credit.

1.12 VOLUME VERSUS PROFIT

The construction industry is growth oriented and volume driven, as illustrated by the popular belief among contractors that ''if you're standing still, you're going backward.'' Contractors pay far too much attention to their own and their competitors' annual volume and even refer to each other in the context of volume and growth: ''There's John Doe, he did $7 million last year'' or ''That's Jim Smith, he went from $5 million last year to $10 million this year.'' Contractors should be more concerned with profits; growth is healthy if measured in profit dollars, not simply in volume.

Emphasizing volume and growth may be fine in a good marketplace but a real problem in a declining one. A contractor's failure to cut overhead and to live with lower volume in a declining marketplace can hurt him. If a contractor is not organizationally and personally prepared to do less annual volume in a declining marketplace, he will need to take greater risks than necessary just to hold his volume. Because he will be trying to get a larger share of a smaller market, he will need to either cut his margin, go farther for his work, seek larger jobs, or do a different type of work. No matter how he plans to hold his volume or grow in a declining market, he will dramatically increase his business risks in an already high-risk industry.

The alternative often is not seen because a volume-driven contractor doesn't consider it an alternative: reduce volume—contract the business instead of contracting for more work. In a declining market one should expect to get less work and to do it at a profit. And the work you do go after should be bid at a profit.

A declining marketplace never put a contractor out of business, but the steps he takes or the business decisions he makes can change his operation and/or affect his profitability to the extent that he begins to lose money. These actions occur preloss or during profitable times; it is the pending drop in volume that precipitates these decisions. If a contractor fully understands the magnitude of the risk and carefully considers the consequences of failure, cutting volume and overhead will be a lot easier to face.

1.13 EMPLOYEE BENEFITS AND COMPENSATION

Another area of exposure for the growing contractor is the establishment of unrealistic employee benefits and bonuses during good years. In a business that is inherently subject to peaks and valleys in market conditions and profitable opportunities, the contractor needs a well-thought-out, sensible compensation approach. Some contractors give out unique and expensive benefits, such as country club memberships and big cars, usually in good or peak years. Because benefits are easy to give and hard to take back, they need to be realistic and affordable over long periods; otherwise they backfire in tough times.

Bonuses should also be considered very carefully. Performance bonuses are quite common in the construction industry and probably overused and abused more than in some industries. A common costly error is in the awarding of bonuses that are not tied to profits. I have seen large bonuses given to field people for just bringing a job in on time—a job that ultimately lost money. A more innocuous problem is awarding bonuses for a portion of a job, like certain trade work, which is properly performed for a profit even when the overall job loses money. From where does the bonus money come? There are far too many jobs in this industry that are salary plus bonus and with the bonus coming every year regardless of the performance of the company.

1.14 BORROWING

Uncontrolled or ill-planned borrowing can also present risks. Uncontrolled borrowing means borrowing without planning or simply borrowing to cover unanticipated operational needs on short notice. When a contractor needs money that he didn't expect to need, he should quickly ask why. Something is not going according to plan. Well-thought-out business plans anticipate the timing and amount of needed capital, and while plans are not always perfect, unexpected borrowing should always be questioned.

Another kind of uncontrolled borrowing occurs when the accounting department is not supervised by a principal of the company. I've seen contractors delegate the use and negotiation of corporate credit to middle management who are not involved with the overall direction and goals of the company. Some of these people can work miracles with banks and can manage to borrow to cover losses for years, preventing the pain of poor performance from being felt in the right places. The planning, if any, for this type of borrowing is a numbers game and not a business plan. Borrowing cannot by itself put a company out of business; in fact, it can save it for a long time. Uncontrolled borrowing is symptomatic of a company out of control. When such a company begins to publish poor financial statements, it may lose control of its destiny to its creditors.

1.15 BUSINESS PLANNING

Perhaps in the building process we are so involved in the planning of the work that we have been distracted from, or have lost sight of, the importance of planning for the business entity.

In a recent informal study of several thousand contractors of all sizes and types throughout the country, less than 10 percent claimed to do any type of corporate planning at all and less than half of them formalized their planning process with anything in writing.

The most common reason given for not planning was size. Believing either that smaller businesses don't need to plan or that planning doesn't matter is a serious business error.

Smaller companies often have limited resources and cannot afford a trial-and-error approach. There are 1.4 million separate contracting companies in the United States, and the largest percentage of them are small businesses. In this tough and competitive industry, many contractors succeed by personally and aggressively driving their business forward.

The importance of effective and efficient comprehensive corporate planning and its impact on the success or failure rate of construction businesses has been understated within the industry. It has been underestimated by contractors for too long.

The construction industry is undergoing some dramatic changes, not the least of which is a realization among constructors that the "tried and true" old ways of running their contracting business are, in some instances, for the old days.

There's always been a painful weeding-out process in the construction industry by which companies that don't keep up fail. Others take their place, and some of these fail.

Many assume that this is simply a fact of life in a high-risk, high-stakes business. After many years of working with distressed and failing contracting businesses, I've determined that the causes in 99 percent of the cases were management decisions alone.

I've described the three primary functions of a construction business: getting the work, doing the work, and accounting for the work. Once your business and technical activities are broken down into these functions, and once your time and energies are budgeted to treat them separately and properly, you should be organized, and this will enable you to properly manage the organization. That refers to managing the day-to-day activities of your business.

The next step is to address the responsibility of top management for the longer range goals of the corporation—the well being and success of the business. Properly managing the important day-to-day marketing, production, and administration areas of the business is not enough to assure success: short-term success in no way implies longer range prosperity.

Without some forecasting and planning, your businesses can be driven in the wrong direction. To carry the driving analogy further, contracting businesses have no reverse gear. If you drive too far in the wrong direction, you cannot simply back up and restart. Once you have committed your resources and company in a certain direction, changing that direction drastically can be difficult, expensive, and maybe too late.

I don't wish to imply that planning is just an effective defensive tool. However, even smaller businesses should embrace corporate planning for that reason alone. In addition to identifying future opportunities and threats to be exploited or avoided, effective planning provides a framework for better decision making throughout the company.

It gives guidance to the managers of a company for making decisions in line with the goals and strategies of top management. It helps prevent piecemeal decisions and provides a forum to test the value judgments of decisions makers within the organization. Perhaps the most significant value of the well-organized planning process is the improvement in communications among all levels of management about goals and objectives, strategies to achieve them, and detailed operational plans.

The planning process properly performed creates a communications network within even the smallest of companies that gets people excited about what's right for the company and how to achieve it.

Planning also addresses an area that is sadly lacking in most small businesses today, that is, the "measurement of success."

Establishing just this fundamental level of corporate planning in smaller contracting businesses has had profound effects on the outlook, attitude, and performance of employees and business owners alike.

One of the greatest selling points for comprehensive corporate planning is that it allows the contractor to simulate the future—on paper. If the simulation doesn't work out, the exercise can be erased and started again. In an exercise, decisions are reversible. Ideas may be tested without committing resources to them or betting the entire company on them. Simulating various business scenarios encourages and permits management to evaluate many alternate courses of action. This could not happen in the real marketplace. Picking the "right" course of action becomes more apparent. There is also the possibility that the larger number of alternatives may produce ideas that would otherwise have been missed. If nothing else, the planning process brings more and better factual information to the table with which management can make decisions.

The mere ability to experiment with different courses of action without actually committing resources encourages the participants in the process to stretch their creative skills in a safe environment.

I hesitate to use the word "model" for fear of this sounding like an MBA thesis. However, models of real-world situations really do give us an opportunity to test different scenarios and their possible consequences.

While no one can predict the future 100 percent, the probability that certain events will have a predictable cause-and-effect relationship is pretty good. The more you know about your business, your marketplace, and your competition, the greater is the likelihood that you can simulate quite accurately the outcome of certain moves. Add this to the fact that the more you plan, the better planner you become, and you should begin to see that you may be able to learn a lot more about the outcome of your decisions in advance than you now do, and that's a formidable tool and a competitive edge.

Corporate planning allows top management to accurately predict new opportunities with greater lead time. With more notice and a predetermined course of action, exploiting new opportunities prior to the competition is much more likely. Another side to this coin and equally important is that being better able to look ahead will reveal threats to the business before they arrive unexpectedly.

The days of assuming a business is helpless in the face of market forces are long gone. Contractors in great numbers today are realizing that their businesses need not react only to marketplace-created booms and busts that have plagued the industry for so long. They are embracing an approach that suggests they can determine their future direction with proper planning. They can be assured that their established objectives are met or they will at least know the exact reasons why not.

CHAPTER 2

Elements of Contractor Failure

2.1 CAPITALIZING ON EXPERIENCE

Understanding the reasons why construction businesses lose money may be the best way we have to prevent unnecessary loss. The investigation and resolution of hundreds of construction company failures have generated a significant bank of knowledge on the subject. The events and decisions that precede the failure of a construction business can be categorized and quantified in order to define the most common causes of these failures.

One of the most interesting phenomena revealed by this study is the fact that the events and decisions that cause or contribute to a construction business failure take place during the company's profitable years. To look for the causes within the bad years when a company is losing money or breaking even is to study the result and not the causes. It is easy to be misled in a study of bad years because losing operations can generate unusual events and decisions even if the contractor is unaware of impending loss.

The events and decisions that precede a construction company failure take place during the one to three profitable years prior to the first year of breaking even or loss.

The events and decisions that precede a construction company failure take place during the one to three profitable years prior to the first year of breaking even or loss. Since many companies struggle through several losing years before failure, the time frame can be from three to six or more years prior to failure.

Having studied the events and decisions that compromised hundreds of companies' difficulties, I have identified ten recurrent and industrywide elements of risk to potential profit or failure. There are two categories of these common elements;

five relate to the company's business strategies or practical considerations and five relate to fiscal or accounting considerations. They are:

1. Increase in project size
2. Unfamiliarity with new geographic area
3. Moving into new types of construction
4. Changes of key personnel
5. Lack of managerial maturity in expanding organizations
6. Poor accounting systems
7. Failure to evaluate project profitability
8. Lack of equipment cost controls
9. Poor billing procedures
10. Transition to or problems with computerized accounting

In this chapter we will explore these briefly, occasionally using very general examples of how these elements affect an organization and its ability to make a profit. In subsequent chapters we will discuss each of these in detail, focusing on identifying and minimizing risks inherent in expanding businesses.

All of these decisions are consciously made, and the events are clearly recognizable and usually appear to be routine business occurrences. Many contractors making decisions concerning growth or the necessity to expand into unfamiliar locations or new types of construction do not see them as risky or dangerous, in fact, with proper planning, most of them aren't. There is no suggestion here that a contractor should fear growth or other change. What is being said is that at least one and usually two or more of these events or decisions preceded the failure of a large number of contractors. There is an inherent danger in these elements, and proper planning and a complete understanding of the risks involved are necessary when encountering them. When two or more of these business changes are undertaken at the same time, they can be lethal.

2.2 INCREASE IN PROJECT SIZE

By far the most common element among contractors who fail is a dramatic increase in the size of their projects. The change to larger projects occurs during profitable years, but problems sometimes develop even before the first of the larger projects is completed. Undertaking larger projects is a natural part of the growth of a construction company; the order of magnitude addressed here, however, is two and three times the previous largest project.

The size of a project relative to the size of the company and to the size of its normal or average projects has a definite and direct relationship to profitability.

When a construction enterprise is operating at a profit doing a certain average-sized project and a certain top size, there is absolutely no reason to believe that it will profit if it takes dramatically larger work.

Almost any construction firm can build a project two or three times larger than it normally does. If a company can construct $1-million road projects or buildings, it can, in all likelihood, construct a $2-, $3-, or $4-million road project or building. As the size increases, so does the strain on the company's resources and technical abilities; however, within this magnitude it can probably get the job done—but can it make a profit?

Making a profit at a job twice the size of a company's previously largest job would be at best unlikely. Making a profit from a job four times greater than the largest ever built would be virtually impossible without both additional resources and a tremendous amount of careful planning. Getting the additional resources required might be possible, but how would a contractor with no background on a project of such magnitude determine what resources would be needed? Without previous experience, how could he carefully plan the work? Contractors who normally do top-sized jobs of $1, $10, or $100 million would be working in an altogether different environment than the one they are equipped for if they took on a $3-, $30-, or 300-million job.

We'll take a $1-million current top size as an example, but the principles hold true for any project. Our contractor's largest project is $1 million; he has two or three major jobs at any given time, say, $600,000, $800,000, and $1 million and probably a number of smaller jobs in the under-$100,000 range. His annual volume is $3 million, and he is generating a comfortable profit margin. When work dries up and backlog approaches zero, he goes after larger and larger projects. He is able to get a $3-million project; in his estimation his problems are over for a while.

In fact, his problems may just be beginning. Let's look at the impact on his organization. Previously his projects took about a year or less to complete. On the average one of his larger projects started about the time another finished, and a third was at its midpoint. On the project near completion he needed to collect considerable retainage, but the one in the middle stages was generating large monthly payments and the one starting up was about to produce some good cash flow through front loading. By handling jobs in sizes he was accustomed to, which normally were in varying stages, he not only had a reasonable cash flow, but also had the time and resources available to look after all of his small jobs and keep them profitable.

Contrast this with a one big $3-million job. At first, the front load is terrific, but the retainage mounts fast and within six or eight months will become a higher amount than he has ever had out on all jobs combined. By the end of the job the amount will be strangling him, and this project will take longer to finalize than anything he has ever done.

While the project is similar to the work he has done, he may be surprised at the level of supervision he is subjected to by the architect or engineer. The municipal, state, and lender inspections may create more red tape than he is used to or than his field staff can effectively handle. Union work rules may be more strictly adhered to on this job besides security and safety requirements.

The larger project, although similar to other jobs the contractor has done, is not within his experience. He can get the job done, but making a profit at it is another story. It would be the same as if a paving contractor who does driveways and parking lots took on an interstate highway project. Building a parking lot and building a highway are similar but certainly not the same.

2.3 UNFAMILIARITY WITH NEW GEOGRAPHIC AREAS

A change from the geographic area in which a contractor normally works is almost as common an element preceding failure as the change in project size. A contractor's primary area may be one county, half a state, or 3, 5, or 50 states. It is that area in which he has normally operated, is comfortable, and has been profitable. While

Building a parking lot and building a highway are similar but not the same. (Courtesy of Caterpillar, Inc.)

there are many good business reasons for a contractor to expand into new geographic areas, such as normal growth, lack of work in his primary area, and special opportunities, the risks must be recognized and planned for. Again, the question is not whether the contractor can build a similar product in a different location. Rather it is whether he can make a profit at it?

A contractor becomes very accustomed to working in his area and can easily assume that his type of work is done the same way everywhere. Yet the differences in customs, methods, procedures, regulations, and labor conditions can be quite significant and expensive if not planned for. I have seen contractors bid outside their areas without knowing in advance that the work would have to be union. While in certain areas of the country it is common to construct underground pipe work practically underwater, in most areas specifications require complete dewatering. In some northeastern states it is almost impossible to keep full crews during the first week of deer-hunting season. There are even some areas where local suppliers will give their best service only to local contractors. Regulatory requirements and inspection will differ greatly from an inner city to the suburbs and may be completely reversed when county lines are crossed.

Without going into geological and weather conditions, there are enough potential differences to cause a prudent contractor to want to make certain he knows what he is getting into when he takes work outside his customary area. Local help, such as a joint-venture partner or new personnel, may be needed to facilitate the project. Compounding the problem, a contractor often takes a distant project that is also much larger than anything he has done in the past because it wouldn't pay to take projects of his normal size so far away.

2.4   MOVING INTO NEW TYPES OF CONSTRUCTION

For a variety of reasons, contractors sometimes change from one type of construction to another or add a new type of work to their existing specialty. Companies may change, for example, from highway work to sewage treatment plants; from heavy industrial to tunnel work; from low rise to high rise; or from high rise office buildings to hospitals.

The need for research and planning before taking a new type of construction work is well recognized by contractors. What is very often drastically underestimated is the entrance cost, the amount paid for the learning period during which an organization adjusts to performing a new type of construction work. Hiring one senior person who knows the new type of work inside out may not be enough. A company may have to complete one or more losing jobs before it can execute a new type of construction profitably. Some companies do not survive this change.

Most contractors are more specialized than they realize. Some bid several types of buildings, for instance, but seem to get mostly one kind. They may call it luck, but it's probably because they are better at bidding and constructing that type of

Most construction projects are highly specialized, and many contractors may be more specialized than they realize. (Courtesy of AGC of America.)

building. Contracting organizations usually start out and remain with types of construction in which they have expertise, and their growth and success are based on the continued perfection of that expertise. Over time they become better able to estimate their kind of work and, therefore, become more competitive at getting it. They also become better at organizing and putting the work in place and become more profitable at it. Being able to plan and execute the construction of a bridge does not mean that you can profitably plan and execute a building.

A more subtle change in type of work is the change from public to private or from private to public sectors. This change, even when the project is a contractor's normal size and in his own area, has cost numerous firms a great deal of money. Again, it certainly can be done with a healthy respect for the differences and risks involved and some planning. Indeed, many companies do both public and private work and have been doing so profitably for years. There is no suggestion here that it shouldn't be done, just a report that many contractors did not recognize any differences in advance and proceeded to bid and produce the work, but not at a profit.

Between public and private work considerable differences exist in

Qualifying for bid lists

the criteria used for selecting winning bids

the amount of collaboration between contractor, owner, and others

the quality of work expected and delivered

the amount of changes assumed to have been allowed for in the bid or to be
   contracted for separately during construction (change orders)

Qualifying for bid lists works differently in the two sectors. In public work, bidders usually need to prequalify with the public body, the state or another agency. However, these lists are open to all contractors, and in most places any contractor can qualify with a little effort. Most start-up general and building contractors achieve their growth within the public sector. Their size of project may be restricted at first, but once they have prequalified, they have a good source of work to bid on because they don't need to "know someone" to bid a public job. This is one of the reasons public jobs usually have many more bidders than private jobs. Most private-sector work, on the other hand, involves select bid lists that are more difficult to get on as owners or architects pick the selected bidders, often in an informal way. Few start-up contractors can find their way onto the better private-sector select bid lists where the number of bidders is usually fewer than on public projects of similar size. The number of bidders on a project statistically affects the number of projects a contractor has to bid to get one. This, in turn, impacts his cost of doing business, which affects his profit margin.

While public bodies are required, generally by law, to award all work to the low bidder, private-sector selection is usually made with as much concern for quality as for price. The public awarding party has no control over the bid list or who gets the work. The parties are often strangers, and the award of projects and the administration of them are at arm's length. A public project is usually administered "by the book." The contractor intends to perform according to the specifications and no more because the quality of work on public projects is usually expected to meet minimally accepted standards. The opposite is true of private work where the awarding party picks the bidders, may or may not open bids publicly, and ends up working with a known or at least preselected contractor. The owner, architect, and contractor are much more likely to collaborate on a private project.

Public work will generally be bid at a lower go-in price than the same private work, and the number of change orders and extras will be greater on public jobs. The reasons are several. The lower price going in on public work allows no leeway to do minor changes at no charge, while on private work, with the team approach and higher price, minor changes in the work are often handled informally with no change orders.

On public projects, change orders often provide the only profit the job will make.

A public-works contractor that makes a profit on most work is often one whose stock in trade is getting paid for any and all changes in the work.

Private work is not bid as tight because it is usually understood by all parties that a fair markup of a contractor's work is expected and numerous nuisance change orders are not.

The contractor for private projects needs to preserve his relationship with the architect, engineer, and owner for future work. He usually builds a reasonable fee and profit into his bid, anticipating the necessity for minor changes, and then sets about building the project including any incidental changes. He earns his fee and profit as a team member.

Very often a contractor who does exclusively public work bids too low on a private job. When he begins to go after extras as he would on public jobs, he runs into problems. The architect and owner not used to this approach genuinely may think they are being mistreated. The contractor has unwittingly created an adversary situation that too often ends in his walking off the job or being terminated.

The differences in public and private work are so little known by the players that constant and avoidable disputes result. The contractor, owner, and architect conduct themselves in what each considers a proper businesslike manner yet the disputes continue. This is probably why the select lists are here to stay.

The quality of work on private projects is usually superior to its public counterpart. When the public-works contractor tries putting minimally standard work into a private job, it gets rejected. When the private-work contractor puts the highest quality work on a public job for which he had to bid low, he loses money. When he doesn't chase every opportunity for extras and change orders, he compounds his loss.

2.5 CHANGES IN KEY PERSONNEL

There are three primary functional areas of a construction business, and each must be adequately managed and supervised in a successful contracting enterprise. The primary functional areas are

estimating and sales (getting the work)

construction operations (doing the work)

administration and accounting (managing the business)

In every successful construction enterprise, a top-level manager is responsible for each of these areas or, in many cases, one person is responsible for all of them or two people share the responsibilities.

If a company is making a profit, it is primarily, if not solely, because of the efforts of these individuals. If one of them leaves, there is by definition no track record of profitability for the new organization. This is a simple reality in business and even more so in the construction business that is so often a closely held small or medium-sized company. You can point to a business with six or eight of the best project managers in the world and tell me, "that's why this company makes money." I'll point to the person who is primarily responsible for construction operations and say, "this company has those six or eight great project managers because of him." You can say the same about two or three key estimators, and I'll say the same about the person primarily responsible for getting the work. Successful companies do not relegate responsibility for primary functional areas of their companies to middle management.

The loss of a profit-making top manager puts a construction company at risk; this risk should not be combined with others. The top management team of a construction enterprise is very small compared to other industries because the labor side of the business is field managed like a subcontract, and some broker contractors subcontract all field work. So the corporate organization is separate and distinct from the field organization. The quality of field management relies solely on the quality of the key person responsible for construction operations. If the key person in charge of construction in an organization leaves, the company is permanently changed and at risk until his replacement proves that he can do the work for a profit as operations provide the entire cash flow for the company.

On the estimating/sales side of a construction business, one key person will be responsible for the firm's bid strategy. This manager will usually take a first-hand part in bid preparations and will determine the final bid. The takeoff and estimating staff may be a great asset to the company, but the top manager put them together, and one person is ultimately responsible for the success or failure of the bidding process. If this person leaves the company, they no longer have a proven organization that can get the work. The areas of administration and accounting are much overlooked and underrated by contractors. If there are two top men in the organization who are responsible for the three primary functional areas of the business, one of them will be stuck with the administration and accounting functions; usually these fall to the person responsible for getting the work because sales and estimating are more of an office function than construction operations are. In some construction organizations it is difficult to determine who is the top manager in charge of administration and accounting because this function is not often recognized as a primary area important to a contractor's success. It is often relegated to middle managers even in medium- and large-sized companies.

This problem is most acute in growing, medium-sized firms. When the business is small, the contractor runs his entire business, including such details as signing the checks. He therefore is close to the accounting side if only by virtue of paying the bills and having a continuous knowledge of his bank balance. If borrowing is required, he is the one who explains it to his banker. Administrative needs are few. The contractor may or may not keep minutes of important meetings, confirm things in writing, or even reply to all of the letters he receives. The contractor is in continuous communication with the relatively few players on his work in progress, and as a result, the impact on his business of the poor paperwork is reduced. As the company grows and the staff increases, administrative and accounting duties are relegated to middle managers. If no principal in a construction firm is responsible for this important primary function, the enterprise is improperly managed. If a dedicated, capable manager who takes personal responsibility for the administrative and accounting functions cannot be identified, the company has a serious problem and is not organized for success. It would be no different from an army marching into battle with no one in charge of its supply line. It's a machine with pieces missing.

Whether this functional area is properly treated by one of the principals or relegated to a manager, if the person ultimately responsible for the company's administration and accounting during profitable years is lost to the organization, it is at risk. Its accounting staff, under new management, has no track record for monitoring the company's progress with accurate fiscal information.

To sum up, one cause of company failure is inadequate replacement of the person responsible for one of the three primary functional areas of the construction enterprise. Typically, the changes in key personnel that contributed to or caused the problem took place while the business was profitable. Adding key people for growth has the same exposures.

2.6 LACK OF MANAGERIAL MATURITY IN EXPANDING ORGANIZATIONS

This element of contractor failure is perhaps the most widespread of all. It is very often found in conjunction with one or more of the other elements; in fact, it may be a contributing cause of all the other errors. Most construction organizations were founded by one man. Contractors who survive the high mortality rate for start-ups usually enter a growth stage. The qualities and abilities required for a contractor to succeed at a small construction business are not necessarily the same as those required for the success of a larger construction business. Confidence and independence, the very traits that make contractors want to be in their own business to begin with, mask for many of them the risks of growth.

Many contractors assume, "If I succeeded at x volume, I'll do twice or three times as well at two or three times volume x." At some point in the growth of every enterprise, however, the organization must change; it must become more sophisticated. At these junctures, more authority must be delegated, more complex systems and procedures will be required, and more sophisticated people may be needed to handle them. It would be nice if these changes evolved slowly over the growth period because they would be less drastic and easier for the contractor to digest. But in the real world that isn't the way it works; the contractor can't hire half of a person or put in half of a new system.

Knowing when and how to make organizational changes becomes an aspect of running the business that tests the true entrepreneurial skills of the contractor in a growing firm. The organizational changes necessitated by growth, particularly major reorganizations, need to be made during successful times to assure continued success.

The contractor who resists change until he has proof of the need for change by having a losing year may have waited too long. Some of the organizational changes required to go from an annual volume of, for example, $1–5 million to $20 million are difficult to recognize and may be difficult for some contractors to accept even if recognized. Delegating responsibility and authority, hiring top managers to supervise long-time associates, friends, or family members, and sharing financial information with more people are a few of the difficult options a growing company may face.

I use the term "managerial maturity" to mean that a contractor's managerial abilities must mature as his business does. He must change from doing everything himself to building an organization that can do everything as well, or even better, than he did. Contractors who are unable or unwilling to change their organizations to deal with their growth effectively must either curtail their growth and level off or face the risk of the business outgrowing its own organization. Attempting to do $20 million worth of business with a $5-million organization is suicidal.

2.7 POOR USE OF ACCOUNTING SYSTEMS

Many contractors spend most of their time getting the work done because they figure if you don't make money there won't be any to count. This may have been a valid approach for contractors in days long gone, before taxes and regulators, before contracts and lawyers, before financial statements and accountants, but it appears that this attitude continues today in varying degrees among many members of the construction community, and it has caused numerous company failures.

The construction industry spends less time and effort on accounting functions than any other industry. Many contractors do not expend much energy on the accounting side of their businesses, may not be familiar with their own bookkeeping procedures, and may not even have selected the accounting systems they use. When this is the case, they are unable to check first-hand the accuracy of the numbers they are given. They are forced to rely on middle managers for the financial information necessary to run a business and for timely recommendations for updating their systems if the business outgrows them. If a contractor does not understand how the numbers are derived and from what data, it is impossible to truly understand the figures, and if this is the case, the contractor is effectively just listening to the opinion of his bookkeeper.

Lack of attention to the appropriateness of the accounting systems in use and to the accuracy of the information generated is lethal to a growing construction business. A stagnant business can usually judge its health from its bank balance, but a growing concern may actually be losing money despite its robust appearance. When a contractor is losing money and his books and records do not indicate it or fail to do so until it's too late to do anything about it, his accounting systems have actually helped put him out of business. Construction is a unique business—fast paced, involving remote sites, and having a lot of money that belongs to others passing through a company's books. A construction company requires systems and procedures that are specially designed to capture this information and efficiently process it to determine accurately what it means. The contractor who does not participate in the selection and updating of his accounting systems loses touch with his business.

Top management is responsible for this critical side of a growing business, and the contractor needs to stay close to it or bring in the qualified senior talent to do it. This does not mean delegating to middle management and keeping an eye on them, or worse, promoting a middle manager without the experience and background for the task. The latter is all too common. A bookkeeper or accountant who has spent all of his career with the same contractor will have no experience to draw on to update the accounting systems when the company doubles and triples its size. With continuing formal education he may be able to handle the job, but he still will have no experience. I am not against staff advancing within a company, but I am against relying upon trial and error.

A great number of contractor failures are caused by poor accounting systems or poorly administered systems that fail to indicate losses in time for something to be done about them. It is the contractor's responsibility to manage this important functional area of his business, and he cannot let it outgrow him.

2.8 FAILURE TO EVALUATE PROJECT PROFITABILITY

Do you know with absolute certainty if you're making money or losing money? This is not a facetious question. I have dealt with distressed contractors who were shocked to learn that they hadn't made a profit in years. Their independent accountants were also shocked. Knowing with any certainty whether or not each construction project is making a profit or generating a loss and how much is an extremely difficult and sophisticated process.

A construction enterprise, whether subcontracting or general, large or small, labor intensive or broker, turns over a great deal of money in comparison to what it keeps. Most firms are dealing with a 2–4 percent profit; when the occasional losing job is averaged in over a five-year period, profit may be lower than that. Profit, when discussed in this context, always means the amount earned at the end of the fiscal year: not one job's profit or gross profit before general and administrative expenses, but the actual profit as reported on a financial statement prepared by an independent accounting firm. The profit a bank or bonding company is most interested in is the one reported by an outside accountant as opposed to a company's internal reports or interim statements.

But who tells the accountant what the profit is? The company's books and records. And who tells the books and records what the numbers are? You do.

This concept is so elementary that it is generally overlooked and more often just not believed, particularly by the outside accountants. The reason is simple. The collection, organization, and presentation of a construction company's financial information is so complicated that most people, particularly those who do this complicated work, begin to revere the results. The sophisticated effort it takes to calculate and develop meaning from the contractor's financial data is so great that the results are seldom questioned. The means justify the ends.

The contractor captures and provides the basic data and is ultimately responsible for its accuracy because it is extremely difficult, if not impossible, for his outside accountants to check or verify the accuracy of this information.

This isn't true for all industries. For instance, in a manufacturing business producing widgets, the outside accountants can verify the company's profit figures for the manufacturing process as long as at least one widget goes through the cycle during the fiscal year. That is, if the raw material is bought and stored and the widget is made, warehoused, shipped, and sold, each step and the costs associated with it can be spot checked and verified from source documents.

In the construction industry, too few projects are bid, awarded, and completed conveniently within a given fiscal year to allow for expedient verification of the base data. To further complicate the verification process, the jobs are spread out and sometimes remote, materials stored and in transit are almost impossible to inventory, and the subcontractor's stage or percentage of completion defies analysis.

To say that the independent accountant needs to rely almost entirely on the contractor for percentage-of-completion information is an understatement. As stated

Construction projects are complex, and it is extremely difficult to determine the exact percentage of completion at any point in time. (Courtesy of AGC of America.)

earlier, a contractor deals in large amounts of money that only pass through his hands. The problem is that all of it has to be accounted for in order to determine whether there is a profit.

To do this with any accuracy at all, a contractor needs to evaluate each job totally, separately, and continuously. And if he doesn't, no one can go back and do it later.

The number of contractors who bunch all jobs together for accounting purposes is astounding. Contractors who simply record all collections and pay all bills without regard for project-level record keeping are not the only ones who fail to evaluate contract profitability. Many contractors keep extremely accurate records concerning the receiving and the disbursement of funds by project; they call this cost control, but it is simply cash flow recording. They can tell you how much they have received and how much they have paid out on a project—and that's it. They can't say with any certainty what they owe, and they pay no attention to what it will cost them to complete the projects. They are not accounting for the profitability of each project.

Methods of evaluating contract profitability are not addressed here. The subject is the danger of not doing it. A contractor who records all costs accurately by job may eventually know if a job produced a profit or a loss, but in the interim he must rely on the percentage-of-completion method to ascertain the company's financial position. Even with the best of intentions the contractor or his staff cannot determine except for the smallest jobs the percentage of completion of a project with any accuracy at all.

Ongoing construction projects just can't be measured accurately. The various accounting formulas generate at best rough approximations of project percentage completed. In an industry that may be dealing with profit margins as low as 2 or 3 percent, a marginal error in the guess concerning the percentage of completion of each contract in progress that makes up the contractor's current financial position could be disastrous. One cannot assume that potential errors in the percentage guesses would be both high and low and average themselves out. All parties in the construction organization involved in the process of determining the percentage of completion of projects benefit directly or indirectly from a higher percentage. There is a natural and understandable tendency to err on the high side and strong potential for overstating the profitability of the work in progress. To compound the problem, most contractors need to use the same data from the work in progress to bid the next job.

This is not an indictment of the percentage-of-completion method of accounting as employed by independent accounting firms to audit a construction company. It will serve their purpose. It is not, however, an acceptable tool for a contractor to evaluate the profitability of his work in progress on a continuous basis.

He needs correct figures by project no later than monthly, and if he is not getting them, he cannot measure his success. The difference between success and failure in the construction business is only a couple of points.

If a contractor does not establish and maintain systems that allow him to evaluate each contract's profitability monthly, he isn't managing his business. He will never be sure of the accuracy of financial statements because data drawn from the business as a whole instead of by job cannot be properly verified.

2.9 LACK OF EQUIPMENT COST CONTROL

A lack of understanding of equipment costs, which include maintenance, operation, and replacement, has led a number of contractors into difficulties, some operating their businesses under a completely false economy. I suggest that owned equipment costs are project costs just like lease costs and should be charged as such. To do otherwise could create a scenario in which all projects report significant profits, but at the end of the year there is a loss because the real incurred costs of unused equipment are more than the profits from the jobs. This occurs more often than not in a declining-volume situation, but I have seen it in numerous growing companies as well. I have also observed it in old-line, well-established companies that failed to account accurately for replacement costs. The problem is one of failure to take into account the real cost of owned equipment. It can be one of not calculating the real cost correctly or not applying it to the projects as they progress.

This problem is the only common element that affects what I'll call real contractors and not brokers. The broker is insulated against this exposure because he generally doesn't own any equipment, and the risks inherent here do not apply to leased or rented equipment.

The cost of owning and operating equipment is a big part of doing business for a lot of contractors. It is important to recognize, record, and plan for all equipment costs even for a contractor who has a limited number of his own tools, but it is paramount to the contractor whose business depends on the use of expensive equipment.

Let's start with an excavation contractor. When he bids a job, say a certain amount per yard for dirt removal, his on-the-job costs will be limited to labor, fuel, and maintenance; these he can readily estimate. To this he needs to add his equipment costs. If the contractor leases a piece of equipment for this job, he has a known weekly or monthly cost and needs only to calculate productivity to ascertain his unit prices. But if he owns the equipment, he may perceive his costs as different from the comparable leasing costs, and in fact, they may very well be lower or higher than the lease price. The lease value may set the upper limit at which a contractor can competitively estimate his charges but has nothing to do with his actual costs. So a contractor bids his work with an equipment charge in it that probably has nothing to do with the real cost of owning it. This is an acceptable approach because in a competitive marketplace pricing our equipment is one of the tools used to get work.

Doing the work for a profit is the next challenge. Here we find a hidden problem in the equipment-intensive construction enterprise: calculating profit. Each project has short-range profit and long-range profit. The short-range profit takes into account only direct project equipment costs. The long-range profit considers all costs of owning the equipment assigned to the project based on usage. Although charging a high equipment cost to a project because the equipment was not otherwise kept busy may seem inappropriate, these are job costs and need to be applied as such. Some contractors charge these costs to overhead; they may become part of the project cost via summaries. But overhead refers to the general and administrative costs of being in business, and if equipment can be leased, the company does not need to own it in order to be in business. It is a pure project cost, 12 months a year, whether or not it is working.

A common error in calculating the cost of a piece of equipment is dividing the purchase price of the machine by its useful life, as in depreciation. This makes good accounting sense but not good business sense. To stay in business, a contractor will need to replace the equipment after its useful life; replacement naturally will cost more than was originally paid. A contractor who uses purchase price to calculate costs is undercharging each project and deluding himself by not accounting for enough money to replace the equipment. The prudent contractor will annually update equipment cost calculations using a best guess of the replacement cost.

Accounting for idle equipment also affects job profitability. Costs go on whether or not a machine is used. For example, if a project is charged for six months' usage and the machine stands idle for three months before it is used again, the real costs need to be applied somewhere. If they are charged to overhead, you don't get a true picture of job profitability.

It is prudent to build into your equipment costs some downtime, which may vary from 10 percent in fair-weather areas to 40 percent in areas where winter weather is a factor. This needs to be tracked and all excess downtime backcharged as a project cost to the last project on which the equipment worked. Making one project or another look good or bad is not at issue here; getting current profit pictures is. The charge should go to the last project worked because its management knows best when a piece of equipment will not be needed; they should start the transfer process (to its next job) early or suffer the idle-time costs.

A contractor who owns his own equipment can usually enjoy a greater profit from his work than a contractor who leases equipment. He must, however, account for all obvious costs of ownership plus the replacement cost. In addition, he must account for all annual costs of owning each piece of equipment and apportion them among his projects or he may not have the accurate information on project profitability that he needs monthly to properly run his business.

The risk of not accounting for downtime is that a contractor may be losing money and not know it. The risk of not accounting for equipment replacement costs is losing money.

2.10 POOR BILLING PROCEDURES

Contractors are often at risk of their cash flow drying up. Those with weak billing and collection procedures are at a greater risk.

Few industries place as much product in the hands of purchasers or ultimate users as far in advance of payment as the construction industry. The terms of payment, basically never in full, are almost unheard of in other industries. The payment approval process, a scenario that includes a series of sign-offs by numerous semi-interested parties other than the purchaser, inhibits expediency almost to the point of absurdity. Late payment for his work is the accepted norm for the construction contractor and the bane of his existence.

An expanding construction company will undergo a lot of growing pains but few as severe as the cash flow problems caused by late payment. Bigger jobs pay more slowly. Retainages in construction are wrong, antiquated, and a ridiculous way of doing business but are a fact of life for the contractor. Their impact on the contractor's business must always be kept in mind. The mere fact of the contractor's need for collection at the end of a job has provided owners with concessions too numerous to quantify.

The billing problem that causes construction enterprises to lose money starts when the contractor of a growing company becomes too busy to chase his money. When his company is small, he is usually close to his customers and generally has limited credit; these two factors tend to facilitate efficient billing and collection of his monthly earnings. As a contractor's business expands and his projects get larger, he becomes more remote from his customers and has less time to do complicated requisitions himself. The billing process becomes a staff function and sometimes is not given the priority it deserves.

It is commonly accepted among contractors that larger projects pay more slowly, and this may be true in part because contractors accept it. It is surprising that contractors, in general, take such a disinterested attitude toward the entire subject of slow payment. This attitude compounds the problem. Many contractors do not get their requisitions prepared and out until 10 days or more after the date called for in their contracts. I have asked numerous contractors about this and have been told almost universally, "These owners don't pay for 45 or 60 days anyway, so there's really no hurry to get the bill to them." But an owner won't start the payment process until he has the requisition, so the importance of billing on time remains.

All construction contracts state when and how payment is to be made to the contractor, but I have seen few contractors exercise their rights and demand to be paid on time. Contractors, who constantly have their feet held to the fire to perform exactly as called for in their contracts and who have every detail in the specifications shoved down their throats, still put up with owners not paying them within the exact amount of time called for in their contracts.

Preventable late payment is a problem throughout the industry. It necessitates borrowing and interest expense and deprives the business of the use of money to which it is entitled. When a contractor is highly leveraged or overextended, late payments cause critical cash flow problems. Money allows a business to operate, and without it small problems get magnified, projects slow down, and productivity can be affected. This erodes profits and for a marginal operation can be enough to cause real trouble. Running out of cash and credit can happen for a lot of reasons; when the cause is simply the inability to collect for work performed in a timely manner, it's inexcusable.

Ineffective billing and collection procedures can directly or indirectly put a contractor out of business.

2.11 TRANSITION TO OR PROBLEMS WITH COMPUTERIZED ACCOUNTING

Record keeping and its vital importance to a contractor's ability to run his business have been discussed at length. Poor accounting systems and failure to use the information have been identified as very common elements of business failure. The absolute necessity of tracking job profitability separately and monthly has been explained. To do these things, a contractor needs to keep accurate and timely records of thousands of business transactions. He may well be aware of the need to convert to and/or effectively manage electronic record keeping, but he is almost never aware of the risks of inefficient conversion and how to avoid them.

When a contractor starts out in business, he is usually able to cope with his record-keeping requirements not only because of the small size of the business but more importantly because the enterprise is new. A new business has no history with which to maintain or reconcile. Start-up record keeping is clean: zero opening balances, fresh project files, and so on, allowing the small company to have less sophisticated systems. As a business grows, the contractor will face the necessity of introducing electronic data processing into his organization.

The process of converting from a manual system, usually overworked and perhaps failing, to a computerized system can be a terrorizing experience. During this process contractors have lost control of their record keeping altogether and have run their businesses using totally erroneous data. When the situation continued for any length of time, the consequences were disastrous.

Converting to computerized systems, while necessary, has a number of pitfalls and is loaded with risk. If approached with the appropriate trepidation and accomplished with professional help knowledgeable in the construction industry, it can be done with minimal disruption. Conversion is usually accomplished more easily in stages than all at once.

Transferring data poses risks. This process takes a considerable length of time, often months, and inevitably will involve multiple transcription errors. Some can be weeded out with additional months of testing or by trial and error; some will remain and, with luck, will be small enough not to matter or to offset themselves.

The manual system should remain the primary system during the entire transition period, even if this lasts for over a year, and should be abandoned only when the duplicate computer system generates the same results as the manual system for two months in a row. Failure to do this puts a contractor at risk. Most companies get a computerized system going as quickly as possible and use it as the primary system. They use the manual system as a backup to check the electronic one. This is absolutely backward. A company has a history of putting construction in place at a profit as recorded and reported by its manual accounting system. It has no such history using the computerized system. Even if the manual system is overburdened, cumbersome, and running late, it is the method that has worked so far. It must be established that the new system can get the same results as the old using the same raw data before it can be used confidently. Switching over too soon channels too much energy to the new computer system and not enough to the proven system.

Managing a failsafe transition is expensive, and the contractor will be anxious to get the faster, sophisticated information he has been promised. And it always takes longer than projected to install, transcribe, and debug the systems. Most important, the people operating the new equipment are usually new hires who know more about computers than they do about the intricacies of the construction business. For all of these reasons, electronic systems are often relied upon before they are trustworthy.

Converting from manual systems to computers will in all cases be more expensive than anticipated. Conversion will cost as much as two or three times the early estimates and will take much more management time and energy to accomplish than anticipated. I have seen this drain on time and money push construction companies almost out of business; if combined with other business weaknesses, it can easily be the last straw. Considering other companies' experience, these costs can be planned for and scheduled.

There is no suggestion here that growing construction companies should not utilize the most sophisticated tools available to them. On the contrary, it is much more dangerous to wait too long to institute needed electronic data-processing systems. Companies can minimize their risk by beginning the transition well before the manual system breaks down and by allocating resources to allow for a lengthy transition period.

The transition from manual to computerized systems should be approached very carefully. It is an insidious danger to a contractor because it appears so harmless— it's a time bomb. Success and survival depend on a contractor's ability to monitor his business at any given time, and the transition from one system to the other can rob him of that ability so quickly and quietly he won't even know it happened. The cost and distraction from other business activities can be troubling even in a successful transition, and the impact on business and profitability can be crippling.

This is an underrated business risk that has done a great deal of damage within the construction industry with very little notoriety.

The transition from one computer system to another has the same risk of losing data but may be somewhat easier because experienced data-processing people are usually already working for the company on the original system. The time involved and distraction from regular duties will still have a cost in time, dollars, and disruption.

CHAPTER 3

Increase in Project Size

In Chapter 2 I briefly described the common causes of construction business failures. This chapter will concentrate on the first cause and the risks involved in undertaking projects that are much larger than a contractor is used to doing. We will also look at alternatives to taking on much larger projects as well as some cautionary or protective steps to take.

The reason that most contractors take on jobs larger than those previously attempted is that the industry is more volume driven than it should be. If rapid growth and expansion is the only reason for moving to much larger projects, then you should at least understand that there are safer ways to grow and expand. For every business, regardless of the industry, there is a limit to the rate at which it can grow safely. The problem is finding that limit before passing right through it.

3.1 LIMITS OF EXPANSION

Determining the limits of expansion is not easy because there is no formula and very few rules. In fact, there are some highly respected management specialists who don't believe there is such a limit. But there are enough companies that were household words in the industry during their meteoric growth stage that no longer exist to make people at least wonder about growth rates. While critics may point to specific reasons for each of these failures, the fact is that rapid growth itself is dangerous—not always fatal, but always a risk.

Each entrepreneur has limits to his abilities, available resources, and capital. Each organization is capable of doing just so much. During periods of very rapid growth, construction companies are so changed that they really become new, untested organizations with much more work to do. The old organization that was so successful is gone forever. Consequently, growth for the sake of growth is risky in any business, but growing in the construction business by taking on projects two, three, or four times larger than anything done before is far more risky than just taking on more projects of the same size.

3.2 REASONS FOR CHANGE

The more common reason for drastic changes in project size is lack of work. A contractor can run out of work for a lot of reasons—declining market, increased competition, higher interest rates, or local moratoriums. When his backlog is down, larger projects seem a quick way out. If the size jump isn't too great, the risk is probably better than the alternative of completely running out of work. But as the size jump increases, so do the risks. A project twice as large as any previous jobs carries fewer risks than one three, four, or six times larger. Often the reasons for taking on much bigger jobs are frivolous and opportunistic: a good client wanted the contractor to bid on it; the job was right next door to the office; the contractor had an in with the owner; or the worst one of all, the contractor got the last look at the other bids. Actually, the risks are far too great for any of these reasons to make good business sense.

3.3 INCREASED RISKS

Basically, the increased risks involved in drastic changes in project size can be attributed to a lack of experience. If you have an organization with a profitable track record doing projects of a certain size, you cannot simply assume that you can profit from any sized project. One just doesn't follow the other. You might say, "I used to do much smaller jobs than I do now, and I'm still making money." That may be true. But how long did this evolution take? Are you assuming that this can and will continue? At what rate? You may be able to double the size of your projects, but that does not mean you can double it again, and again. A geometric growth rate might sound impressive, but it typically turns out disastrous. In fact, it is difficult to know just how large a jump in project size you can attempt at modest risk. We need to know how to evaluate the risks involved before we stake our businesses and futures.

3.4 CASE STUDY

Consider the successful contractor doing $7 million a year in commercial work and multistory condominiums. He's been doing two or three major projects a year and a

The size and complexity of the projects a contractor bids on should be closely aligned to his profitable experience with similar work. Big isn't always better for the organization. (Courtesy of AGC of America.)

lot of smaller work in a two-county area of his state. He's been in business 15 years and has been growing steadily. His average-sized projects and larger projects have also grown. His largest to date has been $2.2 million, but he considers anything over $1 million a major project for him. Almost a third of his volume is small jobs, many under $100,000. Profits have been good and development in his area is on the upswing.

Then an out-of-state developer announced plans to put up a luxury condominium in the area; the contractor's estimator sent for the plans. When they saw the size of the project, they almost sent the plans back but hesitated. The project was large for them, about $4 million, but as far as they could tell, they were the only local contractor bidding the job. The other bidders were larger contractors from out of the area or out of state. They felt they had a real competitive advantage and decided to bid the job. They had three major projects underway at the time, but only one was bonded. Since it was almost complete, they were able to get approval for the bid bond.

The design was first class all the way, and as the bid date neared, it became obvious that the project was closer to $6 million than the $4 million they had guessed. They had some difficulty getting their prices together since some of the specialty items were from distant suppliers and sources they had not dealt with before and because the size of the electrical and mechanical work on the project precluded most of the subcontractors with whom they were used to dealing. This created some difficulty for the estimator because he was dealing with strangers on some very sophisticated systems and controls. He had to be sure that everything was included but not duplicated. Finally, they got the price together but, in doing so, had to forgo bidding several small projects that came along. There was a last-minute snag on the bonding, but a hastily arranged meeting overcame that. The price went in at $5.8 million.

3.5 STAFFING REQUIREMENTS

The contractor had considered the requirement that the project have both a full-time project manager and a field superintendent preapproved by the owner. He thought he would not need both at the start-up or for three or four months while the project got rolling. However, when he got the job, the owner's full-time field representative insisted that both positions be filled from the first day on the site. The contractor had three key field men running his three major projects, one of which was nearly completed. He put his best superintendent on the job as project manager and another as superintendent. This left one man to run the two major projects and finish up the third. The contractor figured he would help out also. When the project manager began to lay out the new job for excavation, he was reminded by the owner's representative that the specifications called for a licensed surveyor. That was the first indication that they were in an entirely new game.

Once underway, however, the project moved along nicely, although both the contractor and his estimator had to spend what seemed to them an inordinate amount of time at site meetings, updating schedules, and reviewing shop drawings and submittals. The shop drawings were a particular problem because, in the past, the contractor and the estimator had never been required to sign off on shop drawings. They reasoned, correctly, that if they were going to sign off on them, they had better review them carefully. It took a lot of time, and they eventually put on a draftsman to handle the preliminary review and coordinate the submissions. The contractor also hired a project engineer as required by the specifications. But he had anticipated that. He did not believe he would ever need to live up to the letter of the specifications when it came to the emergency and first-aid requirements but could put up little argument when he was required to do so.

3.6 CASH FLOW PROBLEMS

The payments on the project were very slow, and the contractor became frustrated when he was unable to get through to anyone in authority at the developer's home office to help him. Eventually, after using up all of his line of credit to keep moving, he threatened to stop the job if he weren't paid quicker.

He was invited to the developer's home office, and over an elaborate two and one-half hour lunch, a senior member of the firm (who happened to be an attorney) explained that they would do their best to pay him as they paid everyone in their normal course of business. The developer made it clear, however, that another threat to stop the job would result in the contractor's termination from the project. They weren't happy with his progress and hoped he would expedite the work to avoid any disputes over losses the owner might incur. The contractor went back and managed to arrange a little larger line of credit and then pushed on with the work.

The contractor's two key men seemed overworked and were somewhat apprehensive. As the job progressed, the level of activity and number of tradesmen were more than they were used to or with which they were comfortable. The constant presence of the owner's representative and his staff (which had grown to three people by the height of the project) consumed a lot of the project manager's time.

The figures on the project weren't looking very good, and the contractor was pushing the project manager pretty hard about them. There was plenty of other work in the area, and the project manager was genuinely apologetic when he quit. He explained that he simply couldn't handle the pressure. There was no good replacement for him from within the company and, with so much work in the area, little chance to hire from outside. The superintendent was moved up, and a good foreman was made superintendent. The contractor had some concern about getting the foreman approved as superintendent by the owner's representative. But that turned out not to be a problem. What the contractor didn't realize was that the owner's representative felt he could get more out of a contractor if the field management was not particularly strong.

3.7 LOSSES

By the time the project was two-thirds completed, the contractor knew he had a substantial loss on his hands. He didn't know if his bid was too low because it had not been a public opening. He was not sure if his bank would extend his line of credit further, and his cash flow problems were mounting. His problems were compounded because the two major projects that were underway when he started the big one had been completed, and both had gone poorly. He knew that the poor finish was the result of taking his best men off the jobs. He had expected one superintendent to do both and close out a third project at the same time. He hadn't been able to help the one overworked superintendent. He had to spend all his time at meetings and solving problems on the big job.

3.8 RUNNING OUT OF WORK

About this time, the contractor came to another realization. He was in the middle of a construction boom in his area but was running out of work. He was doing fewer small jobs than ever because his estimator didn't have time to bid. They'd let the small ones go. Yet these smaller projects had always been profitable for them and were now sorely missed. The larger job was the real problem now. Just after getting the big job they had a $1.2 million job come out from one of their good clients. They spent much time putting their bid together before they found that they couldn't get a bond for it and had to pass it up. In fact, they were unable to get a bond except for smaller jobs. They needed work badly and were promised additional bonding as soon as their year-end statement was available. Their internal reports revealed exceptionally high receivables from the big job, and the surety was getting nervous.

When the statement came out, it wasn't very good. On the asset side the disproportional receivables made it look worse. Tight money caused the big job to lag. Several subcontractors complained to the owner about nonpayment. The owner put the bonding company on notice. The contractor couldn't pay his bills and eventually was forced out of business.

3.9 ONE BIG JOB

You cannot say with certainty that if the contractor had passed up this one big job he would have been in business today. But given the good business in his area, he certainly didn't need it. This example is particularly poignant for that reason. The contractor was profitable and positioned in a good marketplace. At the time he bid the larger job he didn't need the work. It was quite natural to look at the circumstances of being the only local bidder and see a competitive advantage. The only problem was the size of the job. The fact that his original guess that the job would be $4 million when it turned out to be $6 million should have indicated that he was outside his realm of experience. His biggest job ever was $2.2 million, and he forgot that when he looked at that one the first time he guessed $1.5 to $1.6 million. The $6 million job was in the contractor's own backyard and involved the type of construction he did best. The only thing out of the ordinary was the size. Being out of his realm of experience he could not foresee the impact on his other work, his cash flow, his bonding capacity, or his profitability. He had no way of knowing that the second bidder was only $5.7 million—so that wasn't the problem. He did not realize the great risk in taking construction projects that are substantially larger than anything a contractor has done before.

There are enough similar cases to signal that the risks are very real and high when contractors take on a jumbo job beyond their experience. The trick is to find your own niche in the business; that niche includes the project size you do best.

3.10 ALTERNATIVES

What are the alternatives for a contractor if he is running out of work because he is in a declining marketplace? The hardest alternative to sell in the construction business is to do less work. It just isn't in the nature of most contractors to accept such a notion, but it is a very viable alternative. Cutting back overhead and becoming a smaller business to suit the marketplace is very realistic. If the entire market in your area is soft, then all of the contractors will be looking for work. You will find larger contractors who don't usually compete with you going after your work. You will find it easier to compete for smaller jobs if you likewise move down scale to look for work. You will be bidding within your market niche or one that you used to be in and at far less risk than shooting for larger work. The problem, of course, is that your volume will necessarily drop with the smaller jobs. It would be unrealistic to try to hold volume with a much larger number of smaller jobs because in a soft market there are fewer jobs and more competition. If you intend to get these jobs at a profit or at least at the break-even point, you'll get just a share of them, and your volume will shrink.

Another alternative would be to expand your work area and look for work in your best project size elsewhere. The risks involved in geographic change will be discussed in Chapter 4, but this option should be explored and balanced against the risk of going into a very large project. Unfortunately, most construction businesses aren't very flexible. They aren't set up to expand and compress with the availability of work. At one time I competed with a building contractor who was originally a concrete subcontractor. Every time work slowed down, he would take a few concrete subcontracts to keep him busy. He was flexible. We should be cautious of building large organizations with fixed overheads in a fickle marketplace. A portion of our overhead needs to be flexible, that is, overhead that is easily removed, like short-term leases on some equipment and temporary personnel in administrative positions.

There will be occasions when you will, for whatever reason, decide to undertake a job much larger than anything you have ever done. Hopefully, you will have considered all of the alternatives and weighed the risks involved. If you decide to move forward, develop a course of action and stick to the plan.

3.11 UNDERESTIMATING THE SIZE

The first exposure is underestimating the size of the job before you even begin the takeoff work. Don't scoff at this. It's very common. When we're bidding work considerably larger than we are used to, we tend to relate the work items to the scale of the work to which we are accustomed. This is particularly true on work that is not taken off by units—the setting of equipment, clean-up, and so forth. If you are estimating man-days from your own experience, you may forget or not realize that the equipment is three times as big or the building, road, or bridge is twice as large as those with which you usually work.

I was involved with a treatment plant contractor whose largest previous job was $1 million when he got his next largest, which was $3.3 million. Six months later, he was awarded a $7.6 million job with the next two bidders at $9.2 and $9.9 million. A very extensive analysis of the entire bid was undertaken with the cooperation of the contractor, and no mistakes were uncovered. Yet when the bid was carefully compared to an independent takeoff by a qualified estimator, it was determined, quite to the amazement of the contractor, that almost every separate line item was low. There wasn't any single sizeable mistake, but literally hundreds of separate line items on work sheets were consistently low. Although it was the same kind of work the contractor always did, no one in this contractor's office who worked on the bid had ever worked on a job this size. They simply scaled the entire job down in their minds to coincide with their experience or expectations. In fact, the scale on the drawings for this large sewage treatment plant project, which was about 2 inches thick, was smaller than any scale on which the organization had ever bid. The sheets, of course, were the usual size because the work was photo reduced by the designer, which is not uncommon in this kind of work, and the estimators had properly noted the correct scale in their quantity takeoffs. They simply estimated too low in too many places. The size of the loss put this otherwise successful company out of business. Years of hard work and successful projects were all shot on one job that shouldn't have been attempted in the first place because it was too big for the company. It was outside their range of experience, which, in fact, was extensive. They had been in business for over 40 years.

So the first precaution in considering much larger projects is to carefully review the bidding process to make sure you're not scaling down the project in your mind because you aren't experienced in that job size. The second area of concern involves doing the job once you get it.

Again, don't down scale the job and think you can run it with as few people as you typically would to do your every-day work. Determine how many of your key people this job will tie up and for how long. Then take a look at what other work you have and how that will be affected. Also evaluate the impact of tying up your key people as well as your inability to go after additional work in your normal market. If there is little work around, of course, that won't be a problem. Be sure you've asked yourself whether your key people can really do this job and make a profit. Do they have any experience with this size work? If you are going to hire any new people for the project, keep in mind the risk factors. New people are untested in ability and loyalty. You won't know their strengths and weaknesses in advance. If they don't work out and need to be replaced in midproject, you have another set of problems to face.

3.12 NEW OWNERS AND RETAINAGE

If you're looking at a project much larger than you've done, is it with an owner you have worked for before or with someone new? It's always important to know something about an owner. Of course, if you're out of work and desperate, that the owner is unknown probably won't change your decision much. But you should at least try to find out something about his payment procedure and reputation in determining the impact of this big job on your cash flow. Take a realistic if not pessimistic look at the length of time the job will take. Figure the retainage amounts and plan for them. If there is a reduction in retainage when the job is 50 percent complete, determine whether this is definite or at the owner's option and dependent on owner satisfaction.

You cannot merely assume a retainage reduction. Determine the effect on you if you don't get it. If retainage is payable only after final acceptance, you should take a hard look at how long it will really take to get it even if you are dealing with a friendly owner. This planning should take place before the bid ever goes in. Be realistic about how long it will take to get your retainage and the effect on your cash flow.

3.13 APPORTIONING YOUR TIME

Once you have the job, you will naturally want to give it the time it deserves as it will be a significant part of your volume. But you also need to look carefully at apportioning your time and not forgetting the other work on hand. Your other projects may be smaller, but they have always been your bread and butter and may be so again. You don't want to lose the profits they should show, and you may not be able to afford the losses they could generate from lack of attention. In dealing with the owner's representatives, designers, and inspectors, you should keep in mind that they will probably be more used to this project size than you. If you get over your head technically, get help.

I've spent more time in this chapter cautioning against taking on projects substantially larger than you've already done than I have on how to do them. This is quite deliberate because the risks are so great that they should be avoided if at all possible. If you take all the precautions suggested here, there is still no guarantee that you will succeed. I don't believe that a five-story building is like five one-story buildings any more than one $5 million job is like five $1 million jobs. Experience in one project size does not prepare us for similar projects three times larger. Learn first how to crawl, then how to walk, then how to run, and finally how to fly. Leave out a step and you may have to learn how to fall, from a considerable height.

Change in Geographic Location

The business risk in moving from working in a known geographic area to an unknown one is very real and can be evaluated. After evaluating the risks, the contractor can decide to take the step with a prudent amount of caution and planning or to look for alternatives. Essentially, the risk involved in geographic change is real and significant, and changes in project location should be considered important business events and not just "business as usual."

4.1 DEFINING "NORMAL AREA"

Business as usual for a construction contractor is bidding, getting, and producing work in or near where he has so done before at a profit. There is no guarantee that doing work in his normal area will produce a profit anymore than there is a profit guarantee in doing work elsewhere. There is, however, reasonable profit likelihood based on experience. If he has made a profit, perhaps even built a business on doing work in a certain area, then continuing work in the same area involves only a modest risk factor. Contrast this with taking the first job away from the normal area of operations. By "normal area" I don't necessarily mean that a distance of 10 miles or a thousand miles is outside a contractor's normal area. You know what your area is and what it isn't. For someone who does residential remodeling in the suburbs, taking a job 5 miles across the river in the inner city may be out of his area.

The distance varies from one contractor to another, but the premise remains the same: leaving your typical work area to contract elsewhere means, by definition, working in a place where you have no experience contracting for a profit. There is no suggestion here that the job won't get done, but there is a question about profit.

Local conditions are an important element in a contractor's costs, and working in unfamiliar areas can produce some unanticipated conditions. (Courtesy of Caterpillar, Inc.)

While you may, in fact, clear a profit, there are greater risks involved than working at home.

4.2 REASONS FOR CHANGING GEOGRAPHIC AREA

There are many good reasons for expanding a construction business geographically—a desire for growth, the shrinking or drying up of local markets, or the opportunity to follow customers or designers. The question of whether to spread out geographically should be considered very carefully and the associated risks measured and addressed. I have known numerous contractors who jumped vast distances without so much as a second thought. I've seen contractors jump two or three states because they had an opportunity to bid a particular job. Some even proceeded to expand geographically as if it were the most natural thing in the world. Many have made the leap with poor or even disastrous results.

4.3 REGIONAL OFFICE CASE STUDY

Consider a contractor in business for over 20 years getting into his first sizable commercial job—a $700,000 strip shopping center in his home town. He puts a lot of his time and energy into the bid and then into the job and comes out with a fair profit. During the next two years the same customer awards three more centers, and this contractor gets two of them. His normal work area is half of his state. One of the projects is at the edge of his area and the other is in the next state. Each one is over $1.5 million. A year later the out-of-state center is almost completed, and the contractor has landed a lot of the tenant improvements. He decides to lease a small part of the tenant space to open a regional office. He hires a well-recommended, local project manager and takes on a few small jobs in the area from the temporary project office.

The shopping center job was a profit disappointment, but the job had some rock problems early on and weather was a problem later. However, the contractor thought the job might break even as a result of the tenant improvement work. One of his home-office estimators who loved to fish in the area was willing to relocate with his family. After one year the the temporary regional office had gotten only $400,000 in work. They were given one more year to turn it around. The second year they did $3 million in volume. Everyone was happy. This company didn't survive the fourth year.

4.4 WHAT REALLY HAPPENED?

Let's look a little closer at what happened. The first shopping center was a big job for this contractor, and he gave it his personal attention. It was at home, and it made money. The next two were much larger and farther away with one in another state where he had never worked. The only thing the jobs had in common was the developer. The contractor had had one good experience with this developer, but when a job goes well and makes a profit, the experience is usually good for everyone. At any rate, it was clear from the beginning that these jobs weren't going to get as much personal attention as the previous one. Yet no one, not even the contractor, ever correlated concerted personal attention with profits. Getting these two big jobs at nearly the same time, however, was cause for celebration, not pessimism. The decision to hire locally for the out-of-state job was designed to keep expenses down, to deal with the new (local) conditions, and the nearly all-new subcontractors. The project manager came highly recommended by some good people in the area. However, the company didn't know the project manager; he was untested. The rock and weather problems could have happened to anyone. No one even wondered whether or not the local bidders had anything in their bids for these items.

The new project manager had done some contracting on his own. He said he could get some smaller jobs while they were there for the larger one. No one considered the fact that the field office didn't have any key company or support personnel to do these smaller jobs. Nor did they think it imprudent to allow distractions from a project that was the biggest the company had ever done. The truth of the matter was that the idea of a regional office was on everyone's mind from the moment they got the job. Even when it was clear that the job had lost money, they fabricated a quick scenario where tenant work would create a break-even position for the entire adventure. No one questioned that home-office overhead costs associated with the project weren't even charged to it. Not only did they fail to consider the cost of bidding, but the costs of the new telex system, phone bills, payroll service, overnight mailing fees, and so on, were ignored as well.

Putting a trusted estimator at the new office was a great idea, and he used the same successful and tested approach he had learned at the home office. But with no market analysis or study done, the problem grew; the contractor simply planned for the same volume. Consequently, $400,000 didn't even cover regional overhead for the first year. The threat of closing the office in an area where the estimator had just settled his family provided sufficient motivation for him to capture more work.

4.5 THE NEED FOR PERSONAL ATTENTION

Similar situations occur so often that it's embarrassing to our industry. Contractors lose sight of how much effort they put into nurturing their companies. They forget how much energy and night work and worry went into developing a successful operation. They put untested people with very little backup in a regional office and expect profitable construction. A regional office is like another construction company. The contractor provides them material resources like money and manpower, and they build. But he can't give them those essential resources, the natural instinct, business sense, drive, and sense of timing that made him successful as a contractor. He can't give himself, and the job can't be done by close associates or remote control. The regional office that succeeds in the construction industry has more than home-office support. It has a leader endowed with most or all of the attributes of the contractor. If he's really good at it, he's hard to keep because he *is* a contractor.

There's nothing wrong with regional offices in a construction business, but there is risk in starting one. Leaving your own area for a single project can be as hazardous as the regional expansion approach because few contractors seem interested in leaving the security and comfort of their own area for a small job. So you find that a contractor who does a single job outside his own area is usually doing one at the top side of his experience range, sometimes two or three times bigger. Changing project size and geographic area combines two risks and compounds the problem. Leaving your own area and working projects within your size range is risky enough.

4.6 OUT-OF-AREA PROJECTS CASE STUDY

Let's look at a utility contractor who had bid work within a 100-mile radius from his office and was asked to bid on a project 250 miles away. The area was remote, and only two other contractors agreed to bid. The design engineers, well known to the contractor, asked him to participate. He didn't need the job and it was at the high end of his scale, but the fact that there were only three bidders made it very tempting. Although he had never run a job so far from his shop and his supply and service lines, he felt he could keep it fairly self-contained and perhaps try some special expediting techniques to shorten the overall time and save on expenses. This would make a neat lump of unexpected profit if he got the job; and if he didn't, no loss.

He took the job off in his usual way and, in July, took a day trip to the site with one of his project managers. He didn't see anything unusual about the site conditions. In fact, he felt that the remoteness of this farming town would simplify traffic problems and street closings. He asked in the local coffee shop about labor and was told that he could get 30 or 40 people the next day if he wanted them. He could also get quite favorable rates at a local motel if he took several rooms for an extended period. He priced out the job as if it were in his own backyard, adding the equipment transportation costs to mobilization and all of the costs of keeping his key people on site. Then he added a little extra profit for nuisance value and got the job.

The job was to start 14 days after bid opening. This was a little tight on his available equipment, but he could lease a couple of pieces at home to replace what he sent to the job. At bid time, things looked fine. There also appeared to be some farm equipment in the area that might fill in on grading and seeding, and there seemed to be trucks available locally. There was a snag in the county permits, and one of his people had to go out there three times in August to get it straightened out for the engineer. The problem was that the county engineer had no experience with such large jobs. The delay meant that the job didn't get started until mid-September. The contractor had figured four months start to finish but later decided to work his people and equipment 12 hours seven days a week and to give them every other weekend home. They would make a lot of overtime, and he would save on meals and lodging. Since he wanted to do the job in two months, the late start was no problem.

Everything started out as planned except a couple of tag-along trailers were dearly missed on some existing jobs where equipment needed to be moved. The trailers were tied up moving the equipment to the remote site. Then the distance began to take its toll. Everything took a day or two to get out to the site. Downtime for equipment repair parts and needed supplies became a nightmare. There was another big surprise when it became apparent that the crew and equipment were very unwelcome in the town. The project had been planned by a town administration that had been replaced in the last election by a group that ran on a promise to stop the project. The conservative townspeople had no desire for a sewage collection system and plant with its attendant costs and *mess,* as they put it. They had used septic systems for years. They didn't want the project. The part-time mayor read the specifications from cover to cover and held the contractor's feet to the fire. Street openings had to be protected with flashing lights and flagmen, even on dirt roads. The mayor further insisted that the lights and flagmen be provided until all paving repair had been completed. The flagmen were needed four times longer than anticipated to comply with the letter of the contract and to protect streets with a total traffic pattern of 20 vehicles a day.

I won't go on with the details, but after the design engineers went to bat for the contractors, the town administration threatened to throw the engineers off the project. From this point on there was no salvaging the job. The engineers also wanted overtime pay for all the inspection time caused by the contractor's schedule. Because of the difficulty of getting supplies, there was a lot of overordering. Material ran over by 20 percent. In October, harvest season started and all local labor was lost. Some of it was replaced by borrowing men off existing jobs at home, but there weren't enough to hold the schedule. The specifications said no paving could be done after November 30 and no seeding after October 15. The job was stopped. The contractor left and came back in the spring. Labor costs ran over by 200 percent. The total loss on the project exceeded 70 percent of the total contract price. Final payment took two and a half years to collect.

If this had been the only loss resulting from the job, the contractor might have survived the experience. But this remote job impacted in other, unexpected ways on the home-area work. The contractor's plan was to use his best men on the remote job. The project manager was handling three separate contracts when he was sent out of town. Two of them lost money as a result of the changes in field management. The work was handed over in the middle of the job to various other superintendents because two good superintendents had gone to the remote job. And there is no measuring the additional impact of taking the company's four best equipment operators out of the system for three months. Key mechanics, almost all of the service trucks, and even the delivery man were missing at various times during the period. Productivity and morale were the worse anyone in the company could remember. Six or seven jobs lost money because of such impact. Before bidding the remote job in July the company was six months into the fiscal year and enjoying their usual profit margin. Even by excluding the remote job from the figures, they ended the year with a loss. Including the job made the year a disaster.

I'm not saying that a disaster such as this will happen every time a contractor changes his geographic area of operations. But there is risk in moving into unfamiliar territory. Was the farming town story a preventable problem? Should a contractor check the political climate everywhere he goes? I don't know. What I do know is that when you are in your own backyard, no matter how big that backyard is, you usually know the political climate, or the odds are better that you won't get as many surprises. Distance creates costs. Transportation is expensive but at least it can be calculated.

This remote job was handled differently than the contractor's other work. That is the important point. Distant jobs are different.

Moving equipment and men efficiently from job to job and the quick servicing and maintenance of equipment were this contractor's stock in trade. But even he did not know how important they were to his success and profitability. The synergy of moving men, equipment, and supplies in his confined normal working area was the backbone of his efficiency and profitability. The distant job, although it was bid like any other, actually wasn't like the others. It was a totally new experience.

There are some things you can do to minimize and control the risks of doing work outside your normal area. But all of them put together are not as important as stopping and thinking hard before you take the step. Plenty of contractors have done work out of their area, opened regional offices, and made money. It can be done and has been done. We're talking about real risks. The contractors who succeeded were at enormous risk whether or not they recognized it at the time. If you jump off a bridge and live, it doesn't mean it wasn't risky.

4.7 MEASURING THE RISK

Doing a single project outside a contractor's regular work environment is probably more dangerous than opening a regional office, if only because opening a regional office usually gets a lot more attention. Bidding way from home puts a contractor at a disadvantage to local contractors because subcontractor and supplier quotes are harder to get and difficult, if not impossible, to check for accuracy. The contractor who wanders out of his area to bid often gets the job, but too often he gets it because he's got subcontracting prices that the local contractors wouldn't use because they know them to be incomplete or unreliable. Or there may be local site or labor conditions that the contractor doesn't know about or whose costs he doesn't fully understand.

4.8 CONTROLLING THE RISK

The safest way to expand your normal work area is to begin at the edges of your existing work area. You can test your profitability as you go and find your limits.

The safest way to expand geographically is to do it systematically at the edges of your existing work area.

Perhaps you find that you can move along an interstate corridor, on one side of a river, or up a state line before profits begin to drop off. By expanding this way, you limit the drastic changes that may await you in the totally unfamiliar environment of a more distant location. You can also pull back more easily from your perimeter than from a remote location. Dealing with recalls, guarantees, and maintenance periods required by contract is a lot less expensive at your perimeter than at greater distances. Reentry to the area may also be possible after a pullback and some thorough planning. That way, at least the experience is not a total loss.

Hiring locally always makes sense, but expect some training and familiarization time if the job is to be performed the same way the company has always managed its work and profited. If you find out that they don't do things your way in the new area, then you probably don't belong there.

Relocating a top man for the job from your trusted pool of steady people is probably the safest approach. You get consistency in methods and honesty in reporting. This, of course, will strain your home-territory resources, but a contractor expanding in any manner should realize that this is going to happen. It is one of the risks.

Taking one and only one job in the mid or lower range of your normal job size is also a good safety valve, either at the perimeter or at a more distant location. Too many contractors take a large job way from home, and before the results of that move are in, they take one or two more. This puts too many resources in an untested environment and raises the stakes on an already high risk. It's just not good business to increase your bet before the results are in. If you know a distant job is a greater risk than your basic work, it is simple prudence to watch it closely. Take the pulse of the job regularly. Don't go too far afield if you can't afford the time that the experiment or the job deserves.

An alternative to geographic expansion is to remain in your home area, do your same or lower volume, yet give the projects more attention. The usual result is higher profits without expansion.

4.9 OPENING A REGIONAL OFFICE

The opening of a regional office is a good way to get into planned geographic expansion. It's risky, but it can be done. Opportunistic expansion such as opening a regional office at a big job or just because there is a target of opportunity is more risky simply because it involves less planning time, less time to test the water, less time to change your mind, and no opportunity to change the location.

The absolute key to regional office success is the person you assign to run it. The ideal situation would be for the contractor personally to leave the home office and relocate to the regional office. The home office should be running smoothly and profiting if you're planning to open a regional office. So the talent required at the regional level is the same as that which built the business to begin with. Yet moving to the regional office is impractical for a lot of reasons, the most important of which is that the motherlode requires your attention. Basically, setting up a regional office with just a good construction man or administrator to head it up, based on the assumption that it will just be an extension of the home office and will run by remote control, doesn't work. The assumption is false.

If you have the key man in the right location with appropriate home-office support, you have the basic ingredients for a prudent start-up. And that's exactly what a regional office is—a start-up operation, just like a new contracting entity. As the offshoot of a successful company, it should not, however, go through as long a growth and development period. If you donate enough effort to it, the new regional office can grow without the missteps and mistakes of the parent office. Remember, however, that regional offices should start small and grow. Starting out too big is a problem even with the right ingredients.

In starting regional offices remind yourself that there is the local knowledge factor to contend with even if you have done good research or hired someone locally into the organization. A modest start for at least the first year reduces risk. You should get some first-hand knowledge of the area by bidding and doing some work but not so much that the lesson gets too expensive. You get a feel for the area, and you get time to make necessary adjustments.

Distance is not as great a factor in locating a regional office as it is with the remote single project because in locating a regional office you usually follow the marketplace rather than a target of opportunity. "The closer, the better" is a good rule of thumb, but a good market is more important.

4.10 WITHDRAWAL PLAN

This brings us to the last important risk control item involved in opening a regional office—a withdrawal plan. I have observed the most extraordinary turnaround efforts exerted on a regional office that never profited. Regional offices have inertia; once set in motion, they continue in motion. Sometimes, even after the best planning in the world, business decisions simply don't work out. That is why you need a withdrawal plan. A withdrawal, or escape, plan starts with a determination in advance of how long you will continue the effort if it doesn't succeed. You should determine, in advance, the accomplishments and milestone dates you will use to measure the success of the regional office. You needn't be too optimistic. Making a profit during the first and even second years may be difficult. But profit isn't the only measure of success. You may be trying to penetrate a good market or be in on the ground floor of a developing one. Regardless of the reason you may expect to lose money in the first year or two, you should determine, in advance, how much of a loss you can afford. You need to be quite certain you can afford the cash outflow from the home office. You will also want to look at the impact of a loss on your financial statement and credit. You probably won't want to capitalize the loss as an investment, so the bottom line for the entire business will be affected. Your bank and bonding company should know in advance of your plans, particularly if you feel a loss may develop in the first year or two.

So your plan may include profit early on or a loss for a year or two to get started. Whichever, a definite amount should be predicted in advance, and if losses exceed the amount, the plan should come under close review. You may be able to afford the additional losses and continue. If not, withdraw immediately. More than two modifications in acceptable losses should signal a complete review of the entry strategy and, possibly, the initiation of your withdrawal plan. If your plan for a regional office doesn't work out, withdraw before the impact threatens the home office.

Withdrawal is a strategy seldom even considered in construction industry planning sessions. Yet withdrawal from any strategic business move is a valid, businesslike, and often necessary alternative. The reason it is seldom considered is because so many construction businesses are an extension of the contractor himself, and there appears to be some stigma or ego loss associated with retreat. There's no room in the business world for this kind of thinking, particularly in a high-risk industry like construction. To control risks, you plan; to plan, you use all the alternatives available to you. Withdrawal is an important alternative. You should consider it a possibility in every move you make.

Although you might worry about how you'll look in the marketplace or to your peers if you withdraw, what's the alternative? Does it make sense to alter your plan to read: "We will stay here at any cost in order to look good to our peers"? Can you afford that attitude?

4.11 ESTABLISH MILESTONES

In summary, plan your regional office carefully. Establish milestone indications that, if not met, call for the complete review of the expansion plan. If the expansion proves to generate losses that threaten the home office, implement the withdrawal plan. Retreat from the market in an orderly, businesslike manner, making sure you tie up the loose ends and leave the area with good public relations in case you want to attempt reentry. The truth can be a valuable ally in a withdrawal situation. If the competition was too tough for your regional office to succeed, why not be a class act and compliment your peers on the way out. For example, consider a press release like the following: "We find that the fine contractors in this area are well equipped to serve the construction needs of their clients, and our participation in this marketplace, while welcomed, was not needed to an extent that would justify our continued presence." Whatever your reason for withdrawal, you can usually state it in a way that neutralizes the rumor mill. At the same time, your frankness may gain you the respect a smart business decision deserves.

CHAPTER 5

Change in Type of Construction

A contractor seldom changes entirely from one type of construction work to another. That is, he doesn't stop being a road builder and start building only sewage treatment plants. However, expanding into other types of construction work is quite common. What you must remember is that changing the type of work you do is an important business event with associated risks.

You have probably noticed that this book repeats the following premise: successful contractors typically become successful by doing a certain type of construction of a certain size in a certain area. I describe this as *normal work* or *average projects* or *typical projects,* and the fact remains that success doing one type of construction work is no indication of success in any other type of construction. However, I do not suggest that a contractor should never expand into other types of construction, merely that doing so carries with it certain business risks great enough to have caused major problems to a large number of successful contractors.

5.1 REASONS FOR CHANGES IN TYPE OF WORK

If we look at the reasons for getting into other types of work, we find that the most common is a lack of work in a contractor's marketplace. A close second to this reason for change is planned growth; that is, a decision to expand into another type of construction to hasten the growth of the company. There are also such opportunities as a good client or friend having a job to give out that isn't exactly the contractor's line of work but close enough. Whatever the reason for a change in the type of work, it needs to be approached as an entirely new field of endeavor. The risks of doing a new kind of work are the same as those in starting a new business. If the risks aren't recognized and addressed, there is an even greater danger—that of exposing an existing, successful company.

5.2 LACK OF EXPERIENCE

The risk exposure to the contractor comes from the same lack of experience we discussed in dealing with changes in geographic area and changes in project size. The experience problem here may even be greater because a different type of construction can be a whole new world. What's worse, many contractors don't recognize this and try to do the new type of work in the same way they did their old work.

To get a better feeling for the risks involved, let's look at some of the differences in various types of construction. A large number of road builders turned to the construction of sewage treatment plants when highway work slowed down or stopped in their areas—some with disastrous results. To make such a switch, any prudent contractor would expect a certain learning period or plan a first project to get some experience. The problem is that there is no way of estimating the time or cost required to gain the necessary experience. Circumstances force you to make a move like this without even knowing if you can afford it. What's worse is that to get the work, you have to bid it having no previous experience in the field.

A common way to overcome the bidding problem is to take on a joint-venture partner. But here, too, I have seen a contractor complete one, two, or several projects successfully way and yet not be able to make a profit when he then takes a project on his own. That's because being in a joint venture with someone on a construction project doesn't necessarily teach you all you need to know to do the other's work. In fact, if each partner does his share, you may learn nothing.

The road builder can surely estimate and do the excavation, the finish work on the site, and maybe even the concrete work on a sewage treatment plant job. But he has most likely never built round concrete tanks to close tolerances. He would not have the necessary experience in this kind and amount of pipe work and certainly none with the sophisticated control systems.

If you get over this lack of experience at the bid stage, what do you do during the construction stage? Just coordinating the shop drawings and dealing with the interface and physical space problems for all of the systems are nightmares. There are too many examples of building or road contractors not even being able to finish these kinds of jobs without help, let alone making a profit at it. This is not to say it hasn't been done. But it takes a lot of preparation, planning, and some luck.

5.3 MORE SUBTLE DIFFERENCES

There are differences in types of construction that are more difficult to see. Many building contractors look at all building projects as pretty much the same until they

Some jobs are just more complex than others, and a prudent contractor sticks to what he knows best. (Courtesy of AGC of America.)

take their first hospital job. The lessons learned on the first one are usually severe. You find out that these projects refuse to get done on time, are impossible to schedule, and require you to put 3 feet of mechanical systems in a 2-foot space. The difference between a hospital and a residential high-rise is like the difference between an ice box and a refrigerator. Building one does not prepare you for building the other. Different contractors specialize in each.

5.4 KNOW YOUR SPECIALITY

There is no attempt here to suggest that one type of construction is more difficult than another or to suggest that a contractor needs to be smarter to do one than the other. The position taken is simply that each successful contractor is a specialist. By definition, the type of work he has been successful at is his specialty, and these specialties are far narrower than anyone would expect.

A contractor grows and becomes successful by perfecting the skills and abilities needed for his specialty. He can be counted on to get the nuances and subtleties down pat so that he can do the work better and bid it better than his competition. We often fail to realize how very specialized our field is. We're all constructors with an understanding of the entire industry and how it works and fits together. But when we contract to put work in place in return for a fair profit, we are safe only in our known range of experience.

5.5 DRASTIC CHANGES: OBVIOUS

In this chapter I am not going to use a specific example of drastic changes in type of work because there are just too many combinations. I have seen plumbing contractors take HVAC (heating, ventilating, air-conditioning) contracts, electrical contractors bid and get building contracts, and road builders take dredging contracts. There is a long list of contractors who took work completely out of their field and lost money. There is an even longer list of contractors who took work in their own general field but outside of their field of expertise or specialty and lost money. Some went back to their own type of work. Others didn't survive the experience.

A contractor must determine his specialty. Too many, particularly those growing rapidly, don't know where their expertise really lies or the type of work contributing the most to their success. Almost imperceptible changes in the type of work undertaken can make a big difference in the potential for success.

5.6 CASE STUDY 1

Consider a sewage contractor in business for 10 years and doing about $5 million a year profitably when he had his first losing year. He was losing money on his two largest jobs, which were both within his normal, top-side size range and within his usual work area. The jobs were straightforward sewage jobs, and the contractor could point to a string of similar jobs on which he made money. A closer review of his previous projects, however, revealed something that surprised the contractor. Most of his successful projects in the past had been gravity sewage jobs, including all of his large ones. Although several previous projects had force mains in them, these had not been done nearly as well as the others. There had also been two small force-main projects some years back, one of which broke even and the other of which lost money. The sewage work that the company bid was always either a force-main or gravity system or a combination of the two. The contractor had started and was busy building a successful construction business. He hadn't taken note of the fact that he wasn't very successful at doing force-main work. He had not been successful at bidding it, either. The two big jobs that were losing money were both force-main jobs. His estimator reported that because they had missed several force-main bids, they had lowered their unit prices to get the present work. The problems continued, and the contractor did not survive the two losers at once. This contractor and all his people called the company a sewage contracting firm, not even a utility contracting firm. However, while it may not be a common or even acceptable name, by definition, this firm was a gravity sewage contractor in that for their ten-year history they had never made a profit on force-main work or any other type of construction work other than gravity sewage projects. They didn't know their specialty.

5.7 CASE STUDY 2

Not knowing exactly what their real success was based on also caused serious problems for an old-line company. A very large scale, 50-year-old mechanical contracting firm was operating with three divisions just before the company got into difficulties. They had an HVAC division with their own sheetmetal shop, a mechanical division with their own pipe shop, and a service division that also did small contract jobs and contract maintenance. The service division had grown significantly over the years, but because the other two divisions had the greater volumes, service was generally ignored by top management. The service division also did all of the start-ups, punch-list work, and call backs for the other two divisions. This made sense because the service division had dozens of fully equipped service trucks manned by highly skilled mechanics.

The HVAC and mechanical divisions made larger profits than the service business and were taking larger and larger contracts. Management decided that since the contract work represented 80 percent of the company's volume, the service business would be sold. They determined that the sheetmetal and pipe shops were turning out products that could be bought at the same cost on the open market, and because the sheetmetal shop and pipe shop were located with the service division on expensive real estate, they, too, were sold.

The new streamlined organization now had its estimators and contract managers located in an office park and no shop overhead. They easily replaced the volume and increased it 20 percent the following year. Unfortunately, they never had another profitable year in their remaining business life. Current management had not been around when the company was growing. They understood marketing very well and had inherited good pricing methods. What they didn't realize was that the service company absorbed most of the start-ups, punch lists, and call backs because it did this work for the other divisions at cost, 90 percent of it with lower cost nonunion labor. The managers did not understand that while the sheetmetal and pipe shop produced the work for the divisions at the same cost the company would pay on the open market, they produced it on schedule. More importantly, changes and modifications were handled quickly and problems solved overnight. The company historically took a large share of work in a very competitive market and made a profit because it controlled its work by being its own supplier.

With all the problem solvers gone, the company could no longer produce the work for its former competitive prices. Actually, the problem was that top management did not know the real expertise of the organization or the basis of its past success. They got into trouble doing the same type of work they had been doing, only in a different way. In fact, they had no idea on what their success (specialty) was based.

There are a number of other things that make what appears to be the same type of work actually quite different, different enough to create problems and losses if the contractor lacks the experience to deal with them. Take, for example, the differences between public and private work described in Chapter 2, which have cost a good number of contractors a great deal of money.

5.8 UNION VERSUS OPEN SHOP

Changing from union to open-shop work or vice versa is a culture shock all its own. If problems are anticipated, most contractors can survive the temporary or permanent switch. But it can cost a great deal of money. Open-shop contractors taking on their first union job usually choke on the restrictive work rules. This is particularly true on smaller jobs and is not recommended at any price for small projects. Contractors doing the first open-shop job may have difficulty finding the skilled labor they had grown accustomed to having available. The reasons for going into this switch are varied and sometimes personal, but the business risk should be carefully measured. Perhaps the greatest risk is in the potential distraction from other work that the contractor may have if he is tied up learning a new way to do business.

You may think that the risks involved in making drastic changes in types of construction are perfectly obvious. Yet good contractors make such changes and often with disastrous results. So the risks involved are not always obvious. A contractor wouldn't deliberately risk his entire business on one project. And if the risks from drastic changes are overlooked, it is understandable that the risks involved in more subtle changes are never even considered. Actually, subtle changes within type of construction are nearly as dangerous as drastic changes, but we often undertake them with such confidence and are so surprised when the roof falls in.

The risks in change in size and type of project are compounded when a contractor's largest job ever is with a new type of owner, as in moving from paving for private developers to a public highway project. (Courtesy of Caterpillar, Inc.)

5.9 KNOW THE RISKS

In the construction business you turn over a lot of money, and only a very small portion stays with you. So it is imperative that you understand exactly where your expertise lies and how it works to make money for you. You need to know just what type of work you do best and even which subcategories of work you do better. You can then move forward in that type of work with confidence and reduced risk. Once you have done this, you can decide for yourself how much more risk you want to take on as you expand and grow. If there isn't enough of your type of work around to satisfy your appetite, be very careful about going after work that poses a higher risk for your particular organization.

5.10 VOLUME VERSUS PROFIT ALTERNATIVE

Remember, if your volume levels off or even drops slightly because you are going after and getting only the type of work you are best at, then your *profits are likely to grow without any increase in volume or risk.* The more you do at your specialty, the better you will get at it. Real profit growth as opposed to volume growth comes when an organization levels out and everyone has enough time to devote to the work on hand, including the bidding. When you're not rushed or overworked by growth, you'll have time to make improvements in the work you do. As you fine-tune your skills on your specialty, you will find tremendous profit potential.

5.11 TEST WITH SMALL PROJECT

If you take the time to think about it, you will agree that making more money on less volume with little risk is real growth. If you operate like this for a while before expanding into other types of work or bigger jobs, you will at least have a greater cash reserve and more experience when you take on the greater risks. If you must move into different types of work, evaluate all of the risks first and proceed with caution. Test the water with a smaller job first to minimize your exposure. If it doesn't work, you can withdraw gracefully (in accordance with your planning) and try something else. Never bet the whole company.

CHAPTER 6

Replacing Key Personnel

6.1 IDENTIFYING KEY PEOPLE

The key people in a construction company are easy to find. As I mentioned earlier, there usually isn't more than three of these key people, and for most medium- and smaller-sized construction companies, there are only one or two. This chapter may not be very popular with middle managers (who are a very important part of any organization) as it concentrates on the one, two, or three people without whom the company cannot function. We are talking about the contractor. I include in my definition of a contractor anyone who is personally responsible for one of the three primary functional areas of a construction organization whether or not he owns a piece of the business. We are not talking about all the people in an organization with key man insurance, and we are not including nonworking partners or owners or partners who are in the organization but not personally responsible for one of the three primary functional areas.

6.2 FUNCTIONAL RESPONSIBILITY

Let's review the three areas briefly. Every construction business regardless of size has three primary functional areas. They are estimating and sales (getting the work), construction operations (doing the work), and administration and accounting (managing the business). Everything that is done in a construction organization falls under one of these headings.

In each successful construction organization someone has direct, personal responsibility for each of these functions. Any time I encounter an organizational

Each construction company has key people who are the primary reason for the company's success, and the training of the future leadership is important to the continued success of the organization. (Courtesy of AGC of America.)

problem, I find that it can be traced to a person in charge of one of the functional area. Furthermore, a weakness in any of these areas can single-handedly cause a construction business to fail. The loss of the key person who is personally responsible for just one of these areas can and has caused many companies to flounder.

6.3 PARTNERS

How many of us have seen or heard of successful construction businesses founded and run by two partners who, when they split up, are not nearly as successful as they were together? Very often it takes more than one man at the top for a construction business to flourish because some people are only good at doing their own specialty. If one of our hypothetical partners is excellent at construction operations, he may fail without his former partner to get the work and take care of the office. The opposite is equally true. You can, of course, hire someone to do either of these jobs, but it's difficult, if not impossible, to find someone to do it as conscientiously as a partner. Unfortunately, simply sharing ownership with an employee usually won't transform an employee into a contractor. Most employees who have what it takes to be a contractor will have their own plans for achieving that position.

In partnership breakups, the ideal approach would be to fall back to a business no greater than half the size of what you did together. It's simply a matter of risk control. You have no experience working alone so to attempt to work on the original scale is to assume your partner did nothing that you can't do in your spare time. While few contractors will accept a 50 percent cut in volume under any circumstances, at least no growth for a couple of years would be prudent to see if success continues without one of the key players.

6.4 FOUNDERS AND SUCCESSION

The loss of a construction business founder is always a difficult blow to an organization. As discussed earlier, few construction companies are patterned after any particular model. They are fashioned and shaped by the contractor as he progresses and grows. Since many sizeable construction companies were started by relatively young men, it's not unusual to find 30- and 40-year-old companies still run by the founder. These contractors often struggled early on to find what it was they did best; then over time, they developed a way of doing it that worked for them. As a company grows, it often changes what it does a bit or appears to do it differently. But the founder's values and approach are interwoven in the fiber of the organization. The values and business methodology of the founder are the real reason the company overcame the survival rate for construction start-ups and was able to grow and succeed. After many years and managers, these values are often all but hidden. However, you'll usually find them embedded in the basic approach, thanks to the continued participation of the founder. Unfortunately, these values are often seen by the people who will assume control in the future as roadblocks to the company's further growth or expansion and are often described as antiquated or old-fashioned. Then when the successors take over, they change too much and lose the key to the company's success.

This is not to say that companies can't succeed and prosper after such a transition. It is simply a fact that many companies have not survived when they could have if they'd developed a plan for the transition including a healthy, objective understanding of what made the company a success. Many construction companies owe their success to the strengths of one or two key people at the top and do not survive their departure because of a misconception of what makes the company run.

6.5 INACTIVE FOUNDERS

A less obvious variation on the preceding scenario occurs when a founder becomes inactive in an organization and everything continues smoothly for years until the founder dies. A founder can influence a company long after he becomes inactive. Often the loyalty of long-time employees prevents their leaving the new management as long as the founder is alive. Their influence, therefore, helps keep alive old values in spite of the new management. In other cases the new management hasn't the strength to change things too drastically while the founder remains in the background. Perhaps the saddest scenario of all is when a successful management transition is accomplished by the retired founder only to be destroyed as the result of influence by his estate after his death.

Obviously, when the contractor or active partner is no longer around, the company changes. Yet I have seen large, established businesses approach such a transition with no more planning than a retirement dinner. Some top managers in construction companies and even stockholders believe that a successful enterprise will continue to be successful by inertia. Actually, one to three key people drive a construction company. If one of them is gone, the company is in jeopardy until a replacement is found and tested. Even a replacement with equal skills and talents will lack the key man's contacts and connections as well as his loyalties. These are the intangibles that are lost to the company and part of the reason that key men are so hard to replace. It may take a year or more to determine whether or not the replacement will work. These transitions cannot be underestimated.

6.6 SUCCESSION CASE STUDY

We have discussed the three functional areas, their importance to a construction organization, and the fact that a change in the person primarily responsible for any of them puts the company at risk and should be treated with great caution. Consider a successful, second-generation, heavy, industrial construction company doing about $100 million a year that found itself in serious difficulty after the loss of its chief financial officer (CFO). After a successful transition following the loss of the founder (which included some substantial changes in ownership), the company was financially solid and working in a growth market. The successor chief executive officer (CEO) was a good leader and picked by the founder for his conservative approach to the business. He was also chosen to offset a partner who was a wizard in the field but lacked a good business sense. The partners had built the business together from nothing and had decided years before to bring in a third partner to run the accounting department. While he wasn't given equal ownership right away, they wanted someone with an ownership interest in this critical area because they had some close calls in the past with bad figures during a tough growth period.

Even before the transition the surviving partner had been pushing for the company to buy a firm that did much of the design engineering on jobs that they did. The design firm was not very profitable. Consequently, he thought they could buy it and increase their competitive edge by expediting their own work. While the new CEO liked the idea, their only option was a cash deal. And since their debts were extensive, the accounting partner was violently against the deal. He said that using all available cash and borrowing power even in a good market would leave them at the mercy of their creditors for three or four years because the debt service would consume 75 percent of their current annual profits.

The CEO refused to take the risk, but the pressure to purchase continued, and the issue finally had to be settled by the board of directors. They backed the CEO, but the decision really turned on the presentation of the partner in charge of accounting. He was well respected by all concerned and had an excellent understanding of the business, particularly when it came to cash flow. He could remember both the good and the bad years and understood that with a cyclical, high-risk business like construction, a modest cash reserve and some unused credit were not a luxury but a necessity. He kept constant watch over every financial aspect of their business and reported to the board that while they were financially sound and in a growth market, increased competition was eroding their profit margins. Their primary market was construction of industrial facilities for the auto and steel industries. While there was plenty of work there, road building and commercial construction had slowed, causing an increase in competition from other contractors and a decrease in profit margin.

But this was not the primary reason the board found the purchase too risky. What tipped the balance were graphs showing the impact on the company's cash flow eight years prior when their markets declined following a dip in the national economy and a lengthy steel strike. The company had remained marginally profitable during the period in question, but their cash flow had turned critically negative from suspended jobs where retainage could not be collected and from slow-paying clients. Numerous canceled projects added to their profit problems, but what no one realized was that the chief accountant had saved the day by insisting that overhead be cut immediately. His charts showed that without their reserves of cash and credit at the time, they could have gone under.

Two years after the board rejected the purchase of the design firm, the partner in charge of accounting had a serious illness and was unable to work again. The company underwent another serious succession problem that ended when they hired an accountant from their auditing firm who was very familiar with their finances. Everyone was happy with the selection, and since the marketplace was still strong, the future looked excellent. Six months later the partner in charge of construction again proposed the purchase of the design firm. The firm was still available, and he felt they needed it more than ever to improve their profit margins in the field.

The CEO was again skeptical and again it went to the board of directors for a decision. The partner in charge of construction presented a good argument for the purchase, claiming a minimum 2 percent increase in field productivity based on better design service and faster deliveries of design modifications. The new chief accountant reported that he had studied the figures given to him by the field people as well as the purchase price information from the design firm. He thought they could definitely afford it. When asked if the purchase would tie up all their cash and borrowing power, he reported that it would be a simple matter to increase their line of credit. The purchase was approved on the condition that a certain size increase in the line of credit be negotiated before the purchase. No one asked how the new accountant had made his calculations.

As it turns out, the new accountant had taken the average profits for the company over the previous two years and calculated an increase on the basis of 2 percent across the board. This increase was reported to the board as being more than enough justification to purchase the design company. Consequently, the purchase was approved. No one questioned the fact that the new firm was involved in only one-third of the work that the company did. They didn't know that the calculations presented were based on the profits of the entire company and not just an increase in efficiency on the work associated with the design firm.

After the purchase it was discovered that the design firm was even less profitable than originally thought. Some additional cash had to be put into it. However, the increase in their efficiency did increase profit margins on that work even greater than originally projected. So the purchase was considered a success. Two years later one of their major auto industry clients became insolvent. This loss in work cut the design firm's volume almost in half. At the same time the steel industry was slowing down because of import pressures and reduced car sales. In less than six months the construction company saw the cancellation of three major projects they were to start and had five capital improvement projects underway stopped without notice. Their volume was cut in half, and they were suddenly losing money daily. In spite of valiant attempts to cut overhead and find work in other areas, they could not reverse the trend. Their next financial statement was a disaster, and their cash flow made it impossible to pay their debt service, let alone reduce principal.

Their attempts to renegotiate and restructure their loans failed when the bank wanted a written commitment that all necessary bonding would continue to be available to the company and their bonding company would say only that each project would be looked at separately. Filing for reorganization was being considered when a buyer for the company was found. The value of the partners' stock was practically worthless and the engineering company was given back to the original owners at no cost. A change in one key person in this company had cost them their business because the key person was the only one who really understood the long-range measure of risk. Until he left the company because of serious illness, he prevented them from making the ultimate business mistake—taking any single risk that jeopardizes the entire business.

6.7 NEW MANAGEMENT TEAM

For all intents and purposes, changes in truly key personnel of a successful construction company creates a new company. By definition, you have an existing formally successful construction company with a new management team that has yet to complete its first year of profitable operation. It may do very well, and there is no assumption implied that it won't. Likewise, no assumption can be inferred that it will. Such an untested organization may not be willing to cut volume readily to reduce risk during a year or two of proving itself to itself but should at least not push for growth until the new organization completes at least one year of profitable construction operations.

6.8 ADDING KEY PERSONNEL

The loss of a key person is not the only exposure in this area. As a company grows, it must continually add to its management staff, and in rapid or sustained growth new people are often placed in key roles. As in the loss of a key person, you now have a new, larger, untested organization doing a greater amount of work. Part of the organization, the original group, may have proven itself, but the new larger group has not as a team put its first year of work in place at a profit. The new group may or may not be profitable, but few companies take the time to find out because the growth just continues and growth covers so much that a true test never develops. And if problems develop some years down the road, there will likely be new people or circumstances on which to blame it.

6.9 SUMMARY

The loss or addition of key people creates, by definition, a new and untested management team and puts the company immediately at risk. The risk is often unavoidable but, when recognized, can be controlled by avoiding the assumption that all is well that looks well. Construction is a difficult business at best and requires a unique person or group of persons to construct for a profit, and any change in that group is like starting over. The matching, or team, theory exists here also. People who have succeeded separately or in different groups cannot automatically be assimilated successfully into a new group. Each new grouping has its own set of matching-up problems and needs to prove its ability to work together and make a profit before doubling and redoubling its bet.

CHAPTER 7

Managerial Maturity

7.1 START-UP CONSTRUCTION COMPANIES

The exciting business of building things attracts many new start-up construction companies each year and has been doing so for a long time. It is safe to say not only that the largest construction companies in the country were at one time start-ups, but also that, more often than not, they were started by one person.

Attesting to this is the fact that most construction companies carry the name or initials of the founder. Most founders probably never expected or envisioned that their initial efforts would result in the big and successful enterprises that many of them have. In fact, we know that some of today's biggest firms were started by men and women who couldn't find gainful employment and by individuals who simply wanted to be their own bosses. We know that few of these contractors started out with clearly defined long-range plans to become nationwide or multinational construction companies in a predetermined number of years.

7.2 EVALUATION

So from where do big construction companies come? The answer in most cases is evolution—they evolve from smaller ones. To be sure, most large and medium-sized construction companies today have sophisticated planning strategies with elaborate systems and are managed by businessmen/contractors that are second to none in the world. But that's today. Almost all of these successful enterprises can point to a time in their history when the longest-range plan they had was how to make payroll the following week and the most sophisticated system they used was the back of an envelope. What's more, for each one of these successes there are hundreds of failures.

7.3 MANAGEMENT SKILLS

So why did some start-ups continue and become large companies and so many more fail, and why does this pattern continue today? No doubt luck has some effect, but the best businessmen seem to make their own, so we need to look further. And since so many failed contractors were competent constructors and supervisors, we can eliminate these elements at the beginning.

The only discriminating variable, the only real difference, that I can find between the continuing successful construction businesses and the early and midtime failures is management skill. Management skills are not simply more sophisticated supervisory skills. Different analytical skills and thought processes are involved as well as the patience to sacrifice shortcuts for long-term optimum results. Management skills include a certain amount of vision, so that planning for and even dreaming of the future can take place. Because start-up contractors may not be trained or inherently gifted managers, they will need to develop the mature managerial skills that will enable their companies to grow.

The managers of a growing construction organization will need to continue their professional development to meet future challenges. (Courtesy of AGC of America.)

A new contractor finds out very quickly that there is a lot more to running his business than putting the work in place. He needs to get new work, account for the money, administer all the details associated with ordering the materials, meeting the payroll, insuring the business, and so on. Some start-ups do not continue because the principal never anticipated all of this or just doesn't want to do it. Some principals stay with it and just don't do it. Others do it, but not well. This last group makes up the huge number of start-ups that don't go beyond the first six or eight years in the contracting business, and the single most important reason they don't is a lack of management skills.

7.4 GROWTH PHASES

Construction contractors that get through the early mortality years usually go into a growth phase that often lasts for the life of the company. Few get to the six- or eight-year point and level off. Most construction businesses attempt to grow quite rapidly at this stage in their development. Management skills are one of the ingredients (perhaps the single most important one) in a contractor's formula for success. If a business remains stable in size, it can be managed in the particular way

Many successful contractors participate in trade and professional associations as a continuing education experience by sharing what they have learned, putting their time and talent back into the industry so others can learn from them. (Courtesy of AGC of America.)

that has proven successful for it, but when it grows, so must its management. If a business is expanding even at a modest rate of 10 or 15 percent a year, it will periodically grow out of its own systems and procedures; its management needs will change. If the growth rate is anything above the 15 percent mark, and I've seen contractors double and redouble their growth each year, the need for management development is compounded. The growth process itself needs to be managed. Planning for and handling growth is an important factor in an expanding business, and dealing with changing plans becomes a process that must be managed. Getting more work than planned for can be as much of a problem for a contractor as not getting as much work as counted on.

So management ability is critical not only to the survival and continuance of a start-up contractor, it is critical during any growth period. Good or bad luck will affect success but isn't usually a make-it or break-it factor. Neither is working in a good or bad marketplace or a good or bad labor pool because everyone else lives with the same conditions in a given area.

7.5 LIMIT OF EFFECTIVENESS

Each new growth stage presents an entirely new test. A construction contractor running a fast-growing company often finds it difficult to know when he has reached the limit of his effectiveness. Many medium-sized companies won't yet have anyone in the management team strong enough, sophisticated enough, or objective enough to warn them of a problem. For other contractors, the biggest obstacle is an inability to delegate authority within their management teams. I've seen contractors that own large companies surround themselves with weak top managers because they won't give anyone real authority within the organization; many then wonder why they are working too hard.

The number of contractors who have enjoyed many years of success only to fail because they let their companies get larger than they could effectively manage is staggering. This happens when a contractor's business outstrips his managerial ability or when he isn't able or wise enough to bring into the organization the management talent it needs. It takes a lot of managerial and personal maturity to know and admit one has reached the limit of one's own effectiveness.

7.6 THRESHOLDS

The point at which volume will outstrip management ability cannot be predicted; the threshold is different for each contractor and for each management team. There are no specific limits for an operation of a given size or type. The best judges are the contractor and his management team. Unfortunately, the easiest way to see if you have outstripped your management abilities is over your shoulder. That is, only after you have approached and exceeded your real limits, does it become obvious.

The point here is that it is critically important to know that the exposure exists and to be continually on the lookout for telltale signs that change is needed. An organization that is managed by more than one person has a slight edge in its system of checks and balances. The managers are able to observe and critique one another's methods and results; this tends to bring management shortcomings to the surface faster.

7.7 TELLTALES

The telltale signs of trying to manage more than you can effectively are several. A series of internal gripes or small problems, most often concerning personnel issues, signal possible trouble. Especially disturbing are

complaints from long-time trusted employees

an increased turnover among middle managers and field forces

an increase in owner and engineer complaints

a drop-off in the performance or response of trusted subcontractors

an increase in job site accidents

increased absenteeism

All of these things happen in an organization and will grow proportionately with the growth of the business, but when several of them occur at once and for no apparent reason, something is wrong. When you reach the point of reacting to business pressures and concerns and not acting on new issues and planning, continued growth would be a mistake. Without management improvement, expansion could break the organization.

7.8 CHANGE

When a company doubles or triples in size, it is no longer the same company it was before growth, and it usually cannot be successful with pregrowth management methods. Yet change comes hard for some contractors and for good reasons. One of the more difficult changes is to evolve from the close-knit teamwork and camaraderie of a small company to a mid-sized organization in which close contact is no longer possible. Sometimes old-line, trusted employees can't develop to serve the larger, more complicated company, and they blame the contractor, who must try to accommodate both progress and personal loyalty.

Replacing top management personnel is also problematic. Most construction companies have a limited number of top people to choose from in the succession process, and few choose to go outside their own ranks for new leadership, even if the need arises suddenly. With a closely held corporation, this is even more the case. The new leader will, at best, have worked with or managed under the retiring leader and have no outside or independent experience. The level of managerial maturity of the new leader will be tested under the most difficult of circumstances, and there is risk involved.

7.9   DELEGATION OF AUTHORITY

The true delegation of authority within the construction industry is a problem in and of itself. The very personality traits that cause someone to want to be a contractor, to be their own boss, makes delegation difficult and in some extreme cases impossible. I have seen far too many large construction organizations with layers of vice presidents and managers that would not or could not make even the smallest decision without the contractor's input. The contractor often thinks it's a good system of checks and balances, but he can't be every place at the same time and ends up making every major and minor decision in the company and wondering why he's working so hard. This type of setup is commonplace in the construction industry and is just a mistake waiting to happen. When people are give titles and not authority, they eventually leave or mentally retire. Few will put the required effort into decision making when they know they cannot make a decision on their own.

7.10   TEST OF DELEGATION

The test of true delegation of authority is when a person is allowed to make a mistake. That is, when a person to whom authority is delegated to is eventually allowed to make unsupervised decisions that may or may not be correct—they are allowed to make a mistake. I get protests that suggest why allow a mistake, why let it cost money, why not just allow people to make their own decisions in their area of authority and check them before implementation? That's a valid training procedure, but trainees do not run construction companies. Eventually decision makers have to solo and sink or swim on their own abilities and decisions. Everyone will make mistakes—some will cost money—hopefully all will be learned from. But that is the cost of developing top managers in a construction company who can genuinely take some of the load off the founder or key person and perhaps lead the company some day. People qualified to lead in this business will be contractors, and people qualified to be contractors will not stay long with a company that is unwilling or unable to delegate true authority. This goes equally for relatives, some of whom can't leave the business for family reasons and are forced to endure the ''no-real--

authority'' syndrome until their first binding decision is made after they inherit the company.

The inability to truly delegate has affected the ability of numerous contractors to hold on to key personnel and continues to create succession problems in closely held or family construction firms. Judgment can only be developed by using it.

Achieving managerial maturity doesn't just happen; the need for it must be recognized and its skills learned or hired. Understanding the importance of managerial development to a growing construction company is the first step toward it achievement.

7.11 MANAGERIAL MATURITY CASE STUDY

Consider a several-hundred-million-dollar-a-year building contractor over 50 years old that had several executive vice presidents in addition to a score of other vice presidents. The growth of the company was skillfully managed by the founder for years. He presided over his huge organization with a strong hand, had the ability to keep a lot of balls in the air at one time, and worked long hours. To the casual observer he had a strong management team to back him up and carry part of the load, but on closer scrutiny there was very little true authority among his senior executives. There was loyalty and plenty of hard work and long hours, but most of the serious decisions were made or cleared by the founder. As he aged, he spent less time at the office, but given the size of the company the important decisions could still be handled by him even if they had to be delayed awhile. Because he maintained all of his personal contacts in the political and business community, he remained a critical element in the continued success of the company even after he formally retired.

After full retirement the company was managed by a combination of people who had come up through the field construction side of the company and financial executives. There was considerable friction, and several attempts were made by the board of directors, made up entirely of insiders, to minimize the influence or in several instances get rid of the construction men in senior positions. Although completely retired and in ill health, the founder's influence was strongly felt in the board room of the company even though he didn't attend the meetings. The construction men in senior positions had worked for the founder for many years and spoke proudly of the history and values of the company and had worked hard to continue doing business in a similar manner. The financial people on the board pushed for a lot of changes but with the founder in the background had no desire for a showdown. The company remained profitable during the period, but the division among top management grew. Several years after retirement the founder passed away, and within months of his death at a board meeting, which just made a quorum as several members were at job sites, a management committee was named and charged with the day-to-day operation of the company and the CFO became president. Within a year the entire construction organization was subordinated to the administrative/financial department, and a number of key construction people quit or were fired.

The company expanded rapidly and in less than three years following the founder's death more than doubled in size. The profit picture was not very good, but management attributed this to the cost of growth and predicted it would improve over time. The rapid growth had included the introduction of new computer equipment and systems, and this coupled with the distraction of moving to new corporate headquarters produced an administrative nightmare—data input was weeks late and some was lost. This caused a serious delay in producing a certified financial statement at year end, and when an interim statement six months into the next fiscal year showed operating losses and the previous year's statement wasn't produced yet, their surety became concerned and bonding was restricted. Several months later the previous year's results were determined to be a loss of 8% on volume.

The company operated for six years after the retirement of the founder. The first three years equaled historical performance with the fourth marginal, and because of the time lag, the cause of the company's difficulties were not associated with the loss of the founder. What was not recognized here was that the founder's influence was felt after his complete retirement and until his death in a way that forced the company to continue to do business with the values, goals, and in the manner he had established. At first glance you might say this is a case of change in key personnel as the cause of the failure. They did lose the founder—first to retirement and then death—and subsequently lost several very key construction people. But if we look deeper, we must consider the reality that every company founder will eventually be lost to the organization and that that eventually should not and does not have to put every business that goes through it at risk. A change in key personnel happened here, but it was predictable and not untimely, so the cause of the problem is the lack of preparation for it. In this case the cause was the refusal or inability of the founder to delegate authority. There was no one in the organization that had any real authority prior to the founder's retirement and consequently no one who had really run his own department let alone an entire construction company. Everyone did his own job as he had been taught by the founder, who was consulted on each exception. There was no lack of talent in this company, although the more aggressive managers usually didn't stay past middle management. What the company lacked was initiative. Everyone followed the founder, but they worked for him, not with him. There weren't any self-starters; they were jump-started by the hard-driving founder.

The success of this company can and should be clearly attributed to the aggressive and talented founder, and its failure is as clearly attributable to him. It is each managers responsibility to tutor and prepare his successors, and the owner of a company is no exception. In this case, if the founder didn't know this or couldn't do it, he had outstripped his management ability or lacked the managerial maturity to lead the business to the size to which it had grown. You might say that building and successfully managing a business of that size proves his management ability. But if its continued success relied solely on his continued personal involvement, then was it truly a viable commercial enterprise. It is critical for small business owners to recognize the need to train and prepare their successors if they intend the business to succeed them and to understand that learning to run a business includes the use of real authority and the cost of the mistakes or difficulties that might be associated with it.

7.12 SUMMARY

There exists today numerous successful construction companies that will be at risk when the founder or current leader is no longer with the firm as a result of a lack of realistic successor training. They will not all fail, as the new leadership may have the innate ability to do the job, but they are all at risk because they face the transition without benefit of a trained replacement who has practiced running the company or part of it and has proven his ability and honed his skills on the proving ground of real authority. I've been told that a contractor has too much to do already in a growing construction business to be able to deal with time-consuming successor training. That's just another nail in the coffin of growth because if we can't slow our growth long enough to train people, we're putting the enterprise at unnecessary risk. I've also heard "Let them learn it the way I did," and I agree. But that means let them learn by running a small or start-up operation, and that's fine because the mistakes made are on a smaller scale and can usually be recovered from, and the learning is accelerated by the rapid-fire issues that confront the smaller or start-up company. "Learn it the way I did" doesn't mean taking over a moving train without realistic training. It doesn't mean taking over a company that you have practiced on for 30 years to get it to its current largest size ever, one that even you have to struggle to keep healthy. Anyone who uses this sink-or-swim training method should not be surprised at its randomly selective results. With this in mind I hope the term "managerial maturity" starts to take on real meaning.

CHAPTER 8

Accounting Systems

8.1 TYPES OF SYSTEMS

There are as many ways to handle the books in the construction business as there are contractors. The methods range from numbers jotted on the back of an envelope to the most sophisticated computer systems. Most contractors's accounting systems have evolved from small operations. These accounting systems often include components from a previous system together with components from a newer system for current use and from a third system to prepare for the future. Contractor bookkeeping is often segregated into separate systems. A company might have a manual receivables system designed just to tell at a glance who owes them money, and in the same company there could be a highly sophisticated, computerized payables system to identify creditors, amounts owed, and due dates. It would be a toss-up as to whether this company would have a manual or computerized estimating system.

8.2 WHO IS RESPONSIBLE?

I'm not at all critical of the preceding situation of having more than one system, nor am I recommending a different way of keeping records. I am in favor of the *whatever-works-for-you* method. What is of concern is the lack of responsibility that some contractors show toward the accounting end of their business. Some even refuse to accept responsibility for the accuracy of any of the separate systems in their accounting departments. Basically, a manager is responsible for the entire operation, not just for those functions he happens to enjoy. To abdicate responsibil-

Managing the systems and data is an important function in a growing organization because knowing accurately where you are at any given point is critical to success. (Courtesy of IBM.)

ity for accounting ensures problems. This is not to suggest that the contractor do the accounting work himself. But he must know enough about it and remain involved in the function to be confident of its accuracy.

When an outside accountant puts the numbers together for you, he is getting his information from each of your internal systems. He's not there every day to see that they are handled properly, nor does he have the first-hand knowledge of the day-to-day operations needed to know if items are missing or if the numbers make sense. This is one of the realities hardest for some contractors to understand. Many of them find it much easier to shrug and say, ''What do I pay all these bookkeepers and accountants for?''

Accounting problems arise most often not from the accounting systems, but from the information fed into them. The system is only as accurate as the information on which it works. The information flow from the operations side of the business must be accurate and timely so the system can compile only appropriate data. This is critical to get you the reports necessary to run your business. Accounting must be a hands-on concern of the contractor, large and small alike. The larger contractor may have trusted senior people to do the work, but he must remember where the buck stops.

8.3 ACCOUNTS PAYABLE

Once we accept the responsibility for the accounting function of our business, the next step is to be conscious of the potential for problems or errors in accounting and try to avoid them. *The single most serious problem I have encountered in construction businesses is the understating of payables.* If we don't know what we owe, then we can never know our financial condition. There are two phenomena that lead us to this condition: (a) a lack of established methods and procedures to ensure that information is entered *on time* and (b) the deliberate manipulation of payables.

Payables are a problem right from the start because our invoices aren't dated or received in the same month the charges are incurred. In fact, I like to get away from the word "payables" and simply consider *liabilities,* which means we owe something whether or not we have an invoice for it. We incur a financial liability, for instance, for a load of material the minute it's dumped on our site whether or not we have a purchase order, invoice, or delivery ticket. Further, if the material is delivered on May 31 and the bill is not received until June 15, it is still a May liability. You may not call it a May payable because you had no bill in May, the check wasn't due in May, and you may even say it wasn't used in May. But almost everything delivered to the site in May doesn't get billed until June. That is the problem. *Our payables are out of sync with our actual liabilities almost all of the time.* Consequently, we often owe more than our accounts reveal. And as long as it is small, the discrepancy is acceptable. But the contractor who records all payables as of the date received will always have accounting information that is one month behind. Yet every contractor who prepares his pay requisition on June 3, 5, or 7 for work performed in May records the receivables as a May event, in May sales, and rightfully so. But if you take these events together, you'll understand why I say that payables and receivables are typically out of sync in the average construction business.

8.4 YEAR-END RESULTS

Let's take a look at the impact of this phenomenon when it occurs at the end of the month that represents the close of your fiscal year. You have recorded the actual 12 months of sales for the period, but you don't have all of the last month of liabilities (payables) because some of them are recorded as the next month's payables. Of course, you still show 12 months of payables, but because the timing is out of sync, it is a different 12 months from the 12 months of sales. You are missing some of the liabilities for the last month of the period, but you have included all of the liabilities that were bumped forward from the previous fiscal year. This may not be a problem for a company that does the same volume every year, but for a growing company in which the current year's sales are a little or much larger than the previous year's (and particularly in the last month of the fiscal year), the distortion leads to a gross overstatement of income (profit). You have recorded all of this year's sales but not all of this year's liabilities. Instead, you have substituted part or all of a month of last year's lower expenditures for part or all of a month of this year's higher liabilities.

If a company is not growing, this phenomenon only leads to a small distortion, but in a fast growing company, a large distortion can result. The important fact is that *any* distortion in financial data is simply unacceptable. You need to know if you're making or losing money and how much. I've seen too many cases of this distortion being greater than the profit reported for the year and have observed the shock of too many contractors who, when the distortion was pointed out, realized that what had appeared to be a good year was really a loss.

8.5 DISPUTED INVOICES

The proper way to record payables is in the month they were incurred and not when they are dated or received. This includes disputed payables. I've seen hundred-thousand-dollar invoices left off payables because they were returned to the sender with a $1000 disputed amount. Neither the disputed thousand nor the undisputed ninety-nine thousand was recorded as a payable. Some companies do this continually.

When disputed invoices are finally recorded, they are typically recorded as of the current month because all of the previous months books have been closed. So some of these are bound to span the close of an accounting period, causing further distortion. You can say these amounts shouldn't really matter on an overall scale of things, but actually these distortions are always on one side of the ledger. They all tend to overstate profits. If you get several of these at the same time, the distortion can be a significant amount. The main point is that you can't continue to follow procedures that confuse the system and deliver distorted results. There is enough potential for error in what we do without building it into the network.

8.6 CASE STUDY

I was involved with a contractor who was having some serious financial difficulty but whose financial statement showed profits for the 25-year history of the company. He was baffled because cash had run out and debt had increased to his credit limit, all while his financial reports showed steady profits. His annual volume was about $20 million, done mostly between April and October. His fiscal year ended in October, which was usually a high-volume month because he was pushing to finish as much work as possible before the cold weather. For his last two years, September and October had been the biggest profit months and November and December the worst. But nobody ever questioned the fact that a lot of September and October payables never hit the books until November and December. However, this was not the only distortion affecting the company.

I asked the controller to walk me through the payables approval process after the office closed one day so that I could get a better handle on the current payables. We began in the president's office where the incoming mail was put each day. The president would look over all the invoices as his way of keeping up on them and would initial which project manager or superintendent they were to go to for approval. Then he would put them in his out-basket. There were a number of bills there at the time, and I recorded them.

Each day his secretary would distribute them to the in-baskets of the appropriate project managers or superintendents who came in each Saturday to approve or disapprove the invoices, among other duties. I recorded the several piles of invoices in both the in- and out-baskets and remarked that the length of approval time on the invoices averaged about one (sometimes two) weeks. I was told that most of the project managers and superintendents took the invoices back to their job sites for an accurate check of quantities and returned them the next time they were in the office.

Each Thursday these approved invoices were moved to the controller's in-basket where he looked them over as his way of keeping up. This was always done on Thursday because the bookkeeper set aside Thursday to code all approved invoices with the vendor code for input into the computer by the part-time person who came in once a week.

What no one knew was that to avoid being rushed, the data put into the computer was that which was coded the week before. Weeks of delays were built into the accounting system. If there were quantity disputes or if a superintendent got behind in his paperwork, months of delays were built into the system. When I added up all of the invoices that were in the various in and out boxes, the result was $2,167,000 worth of unrecorded payables—many of them over 30 days old and some over 45.

Because of the continuous time lag, this company never had an accurate picture of what it really owed or what its true cash needs were, let alone its profits, if any. Because their business was seasonal and the fiscal year ended at the end of their season, the time lag built into their system gave them a totally false picture. When this was pointed out, there was great debate from their accounting people. They had a computer-generated payables aging report, and the aging hadn't appeared to be all that bad until a short time before I was called in. Yet even the accountants didn't know that the payables were entered by the part-time data processor using the date on which she entered them and not the date of the invoice and that, therefore, they were being aged against an erroneous date. If any of this sounds vaguely familiar, I suggest that it may well be time for you to give some attention to your accounting function.

8.7 PROPER RECORDING OF PAYABLES

The proper way to record payables is upon receipt. They may be recorded as *not approved for payment,* but they must be captured by the system as a potential liability rather than be passed around the office under the assumption that they are not owed until approved. If an invoice should be disputed, disapproved, or changed later, a reversing entry will be made. It is a matter of control and accuracy. For the company to know where it stands at any given point, it needs to know what it owes, and the invoices must be captured by the accounting systems at the earliest possible time. By passing the invoices around before recording them, so much time can be lost as to lose control of payables altogether. Some companies even mail the invoices they receive to their various sites without first recording them. Of course, there's no real problem if you lose an original invoice because eventually the vendor will send you another one. The problem is one of knowing your real liabilities at any given time so you can tell if you're making money *now* and how much or losing money and how much.

8.8 ACCOUNTS RECEIVABLE

In my experience, this problem does not typically afflict the receivables systems—the records of that which people owe you. It appears that all contractors record sales correctly in the month the work was done irrespective of when the requisition or invoice to the customer was prepared or dated. There is, however, a problem with the aging of receivables, which in some cases can cause distortions in the profit picture.

When receivables become uncollectable, they need to be written off. That is, they should be taken off the books as assets and, of course, entered as bad-debt expenses. Consequently, the bottom line will be affected. In tough times, contractors are reluctant to write down old receivables they know they won't collect or old retainages they have all but waived. They are reluctant because of the impact on their financial statements. If you are doing this, manipulating receivables to maintain your credit line, you must make the adjustment in your head and know your real position as opposed to the position on the statement. It's imperative that we know and react to our real financial condition.

It's not good practice, but if for any reason your statement doesn't reflect all of what you know to be the case, then adjust the numbers in your head to reality and manage your business using the right ones. It's best to take write-offs in a timely fashion. Only you know what is timely. You can't expect your accounting department to know when to do it, and the system most people use keeps every receivable on the books unless it's received. You need to look at these lists periodically to manage them for accuracy.

8.9 TIMELY DATA ENTRY

All accounting systems and subsystems are just methods and procedures for capturing information about operations to give you a numerical picture of how you're doing. They don't do anything in and of themselves. They don't produce a product. They just collect data, fold it over in several ways, and give the same data back to you in a usable form. The systems can't improve the data, but they can give you an altogether false picture if you misuse them or confuse them. Entering information out of sequence will generate false pictures and even correct properly entered but very old data, generating incorrect information. The data may be correct, but the way it has been folded over with other correct data from a different period creates a false report that's of no help to you. *If the information is only a little wrong, then it's only a little useless. If it's a lot wrong, it's not only totally useless but downright dangerous.* With this in mind, I would again suggest that every contractor needs to understand the accounting systems he uses in his company and be close enough to the output data to verify monthly that it makes sense as it relates to the month's operations. There are a lot of subtle ways that the timeliness and accuracy of information can be affected on its way to the accounting systems, but most are quite obvious and the whole idea of putting accurate information in on time is not very complicated. It's pretty straightforward. *Use a system that's right for you. Make sure everyone puts the information in correctly. Discipline that decision, and use the information generated.*

8.10 UNCOVERED CHECKS

There are other ways to directly manipulate the systems that are (or should be) obvious to the contractor. Some of these can be very dangerous. One example is the

Information must be put into the accounting systems in a timely manner for it to produce accurate results quickly enough to be useful. (Courtesy of IBM.)

practice of writing checks that can't be covered by current balances and are sent out later when there are sufficient funds. Sometimes this is done in anticipation of collections and sometimes (I've been told) for convenience. The problem, again, is bad data. Payables are reduced or wiped out when the check is written. But if there's no cash to debit, the cash goes into the negative. But because most systems don't handle negative cash or because most bookkeepers and accountants don't like to deal in negative cash, the cash account is simply not totaled until enough money comes in to cover the checks, at which time the checks are sent out.

What you have for that period of time is some misinformation in your system. Your records are wrong. When the money comes in and the checks are sent out, the wrong is righted—until next time. It's bad practice and can lead to problems. One such problem occurs when the expected money doesn't come in or not enough comes in, so only some of the checks get mailed. In the meantime, other checks need to be written for day-to-day needs, and the ones that aren't sent get old. Instead of being voided, they are sent when money is available and while new checks are being written. Then these new checks are not sent until money comes in, then only part of those are sent, and so on. It doesn't take long before the finances are totally out of control. Unfortunately, this practice becomes a habit, and I've seen a lot of companies continuously write and sign checks that aren't covered. Some even have an elaborate system to determine which checks to release as certain funds arrive. The problem, of course, is that their books and records are always wrong, and so are their reports. Of course, they could always void all the checks and reverse all the entries to get an accurate picture. But that's a lot of work and most don't do it. Some don't even bother to clear the books at the end of their fiscal year. The methods they use to get this practice past their outside accountants vary. Suffice it to say, this practice is worse than bad form and can lead to a real loss of control of the company's finances and of the company.

8.11 SLOWING PAYABLES

The most blatant manipulation of accounting systems is the deliberate slowing of payables as the end of the fiscal year approaches. This has the effect of producing a false or inflated profit on the statement and is not only bad business, but leads to a real loss of financial control. The payables need to be fed back into the system at a later date, of course. So for at least several months, if not for the whole year, everything the records show is wrong. One can't manage a business with such erroneous data. But then people who resort to manipulating the data in this way have ceased to run their business. Their business is running them.

8.12 SUMMARY

If you don't have good accounting systems that give you accurate and timely reports, you need to get them to manage your business properly. If you have the systems but don't use the reports on at least a monthly basis, you're not managing your business safely. You need to know enough about the systems to feel comfortable in judging that the output numbers make sense in the context of current operations. The collection, compilation, and calculation of construction accounting data is so complex that the people who do the work begin to revere the results. That is, they begin to respect the resultant numbers too often without checking or in some cases not even knowing if they make sense or fit with current trends or expectations. The contractor on the other hand, who knows enough about where the numbers come from, should question results that don't fit normal patterns or his expected results.

CHAPTER 9

Evaluating Contract Profitability

9.1 DIRECT INVOLVEMENT

The construction contractor handles a lot of money that isn't his. We get to keep only the profit, which we all agree is too small for the risk (though you could never convince your employees or the general public that it's small at all). Most medium-sized and larger contracting firms would be happy to keep 2 or 3 percent of sales if they could do it. The basic problem most contractors have with evaluating contract profitability is that they leave the accounting to the bookkeepers and accountants. That's not to say that bookkeepers and accountants can't do accounting work but rather that too many contractors avoid accounting like the plague. They don't like it. They don't do it.

Contractors who come up through the ranks or from engineering schools may never have had even a basic accounting course. They can do math and geometry but they don't want to deal with debits and credits. Yet, when a contractor doesn't participate in the accounting function of the business or has a limited interest, the bookkeepers and accountants are left to collect the information and develop reports and statements without the help of people who can say from everyday knowledge of operations whether or not the compilations of numbers in the various categories make sense. Some contractors have so little use for the accounting side of their business that they even resist passing along all of the real data.

9.2 ACCOUNTING FOR PROFIT

The primary reason for being or staying in business is making a profit. Even if you're in the construction business for other reasons, you won't be able to stay long without making a profit or at least breaking even. The difference between a profit or loss on some jobs is not much and could involve only a couple of good breaks or bad ones. But if making a profit on a job is difficult, the only thing harder is accounting for it. Capturing all of the data accurately and in a timely fashion and using it properly to track interim profit or loss and then final profit or loss in one of the most difficult and misunderstood processes in the construction business.

You can say profit equals sale price minus costs. That's it. And if the job lasts about a week and everyone bills you properly and on time, and if you don't forget the fringe benefits or company-owned equipment, and if you eventually collect retainage and remember to back in the call backs and guarantee work, then it's easy. For construction companies whose work lasts a little longer, however, it starts to get complicated. It begins to involve all of the various accounting systems in the organization to determine if we're making money or losing it.

9.3 SELECTION OF SYSTEMS

If we are going to account for all of the money that passes through our hands (if only to know how much we get to keep), we need to participate directly in the selection and use of our accounting and bookkeeping systems. They must make sense to us and fill our needs as a tool to run the business. They must be accurate and we have to make sure they *are* accurate. It is critical for the contractor himself to participate in the accounting process. He need not do the actual work, but he must check on the system if the numbers in the broad categories and subcategories are to make sense from month to month. He can't just come in once a year and do this. He must remain involved in the accounting function each month for it to make sense. I've seen overhead costs reported and accepted that were 50 percent of the previous month's, and no one even sensed that there must be an error. I've watched sales reported improperly, and no one even realized that no billing went out on one of the largest projects that month. A contractor who watches the numbers side of his business continuously knows before his accounting department does the approximate month's volume and costs. And he should. The bookkeepers and accountants can only add up that which is given to them in the normal course of business and put these figures into the proper categories to develop reports. But if someone doesn't bill you for their subcontracting work for last month, only you (not the bookkeeper) know the work was done and the cost was incurred.

I've had contractors tell me that as their business gets larger, they can no longer pay attention to the numbers each month. Yet to manage properly, you need to know on a regular basis whether the information you are getting makes sense. You get reports. But are all the costs incurred included? Does the report show a 20 percent gross profit when you know that's just not the case? You have to make sure it makes sense. Making sure the accounts make sense may mean acquiring better systems and a controller instead of a bookkeeper, or it may involve (depending on

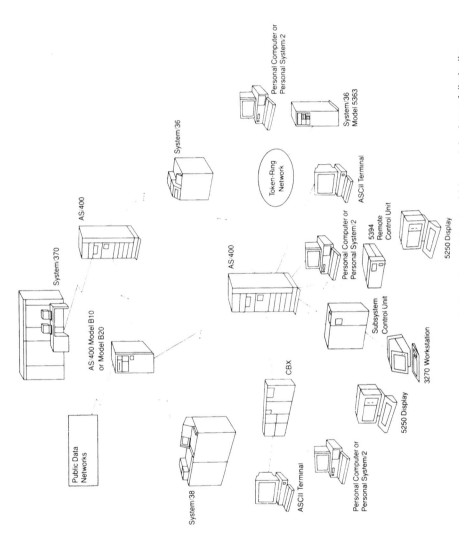

The systems and methods to collect and use data should be matched carefully to the company's current and future needs. (Courtesy of IBM.)

Public Data Networks

System/370

AS/400

AS/400 Model B10 or Model B20

System/36

Personal Computer or Personal System/2

System/36 Model 5363

Token-Ring Network

ASCII Terminal

AS/400

Personal Computer or Personal System/2

5394 Remote Control Unit

5250 Display

Subsystem Control Unit

3270 Workstation

System/38

CBX

5250 Display

ASCII Terminal

Personal Computer or Personal System/2

the size and stage of growth of your company) hiring a qualified vice president of finance instead of a controller. And even the contractor himself can't abdicate the responsibility. However, if he genuinely does not have the time, training, or ability to manage the accounting function, then a real board of directors is necessary as a safety valve. Outside board members always seem to know if the numbers make sense. They're far enough away from the trees to see the forest.

9.4 PERCENTAGE OF COMPLETION

The state of the art as it applies to construction accounting has improved over recent years but continues to be an area of great concern. The primary problem lies in the fact that an independent accountant has no way to verify the accuracy of the figures for percentage of completion that the contractor gives to him for each project. For that matter it is even difficult for the contractor himself to determine if a project is 40 percent complete or 38 percent complete. You can't just stand on a site and say, "Oh yes, this project is 62.5 percent complete." The exposure here should be obvious in an industry where the real profits for any sizable contractor are in the low single digits. An error of 1, 2, or 3 percent in the estimated percentage of work completion at statement time can make all of the rest of the accurate information on a statement erroneous.

We might hope that the different estimates for all projects would average themselves out with some being high and some low. Actually, however, since the people making the estimates have their performance and bonuses on the line, it is perfectly human for them to estimate the percentage of completion on the high side. For a construction company of any size at all, few projects start and end in any one fiscal year. This means that a large part of the income claimed for the year will be based solely on the estimated percentage of completion on all the work except that which closed out during the period. Therefore, we get to claim, as earned income, a portion of the originally estimated profit on the work. If we guess the amount of work completed to be higher than it actually is, we are claiming more profit than we have earned in the period. If a company is growing, it will have more and more work in progress each year, and if the estimates of completion percentage are slightly on the high side, the previous errors in completion estimates will be covered. That is to say, if the percentage of completion of projects is exaggerated in one year, it will probably go unnoticed as long as there is more work in progress the following year.

You might say that eventually it all comes out in the wash because when the project is closed out, the totals are reported. This may be true, but in the construction business you need to know accurately where you are now—if you are making a profit or loss on the present work and how much. One of the greatest exposures a contractor faces is to continue doing the work in a way that doesn't make a profit while thinking he is because the numbers say so. The danger is that no steps are taken to do the work differently, and all go merrily on their way.

9.5 ESTIMATED PROFIT

There's an even more insidious problem with this procedure if the real profitability of each project is not reviewed periodically. It is necessary periodically, and certainly at statement time, to reevaluate the actual project profit against the originally estimated profit. Even if our percentage-of-completion estimate is accurate, we may be overstating our income if the job isn't generating the originally estimated profit. So we need to decide not only what percentage of the profit is earned, but also whether the profit is there or not and in exactly what amount.

9.6 CASE STUDY

I was involved with a contractor who had been in business nine years and whose company had grown to a volume of $70 million in that time. With such rapid growth, each year there were more and larger projects in progress and, of course, in progress at statement time. For each year including his last, this contractor's certified financial statement indicated strong profits and, after nine years, a very substantial net worth.

The fact of the matter was that the company had a negative net worth. The negative worth did not simply result from recent losses or losses on current projects. As the jobs became larger, they also ran longer. Consequently, there were not only more new projects each year, but also more ongoing projects from previous periods at each statement date. The profits from the increasing amount of work, based on the percentage-of-completion method, were substantial and always much greater than the previous year. This meant that a losing job reported as profitable in a previous period was well covered, and the total for all jobs for the current period was a positive and substantial amount.

What had been happening, of course, was that profits were claimed each year for jobs that in subsequent years were reported as a loss because they actually never made money. There were quite a number of these, but with so much growth, the statements always looked good and revealed profits. Finally, however, the cash ran out and a subsequent study of all past and existing projects showed some startling results. The overall picture was a disaster, but what was most interesting was that of all completed projects for the past years, more than 65 percent had lost money. These losses were covered by the percentage-of-completion estimated profits of an always greater amount of work in progress. During this time the ongoing jobs were shown to be earning their originally estimated profits, but many would eventually be reported at a loss. The contractor didn't do anything about his losing projects because he didn't know they had occurred until they were finished.

9.7 COST CONTROL

The percentage-of-completion method is good enough for statement purposes, but the estimated figures aren't good enough to use in managing your business month to month. You must have a real cost control or job costing system that captures all of the project costs and captures them on time so that you know, with reasonable certainty, whether or not a job is making money. And you need to know this in time to do something about it. The timeliness of information here is extremely important. If your accounts payable system records costs in the wrong month as described previously, then it's more than likely that the same misinformation is getting into your cost control system. If you're recording the sales on time and some of the costs a month late, then each job will look great on paper until a month after it's finished.

I have seen many sophisticated, computerized cost control systems with bad data entered into them or good data with the wrong dates. When that happens, these systems become inherently dangerous. Getting the wrong information and relying on it to manage a business is worse than no information at all. This is because almost every potential error and timing problem tends to lose or ignore costs. When the lost costs are ignored, the reported profits increase. *But these profits aren't real.*

Cost control systems do not have to be complicated or computerized. There are many different kinds of systems, and this text does not include recommendations or go into the mechanics of them. What will be discussed is what the system should do and what has to be put into them and when. The difference between job cost systems and cost control systems is that job costs merely collect the amounts spent to do the work to date. I've even seen these systems set up to collect costs for all jobs bunched together so that all costs could be compared to all sales for the month. It should go without saying that each job must stand on its own. We need timely profit and loss information by projects, sometimes even by separate activity, but not as a group. A cost control system collects costs by job in categories that can be tracked against bid or estimates. It actually controls nothing but gives us information about how the project is running against planned or budgeted amounts so that *we* can better control the project.

9.8 TIMELINESS

The key to the success of a cost control system is getting the information in time to do something about it on the job. The information, of course, needs to be accurate to be of any value, but trading off accuracy for timeliness gets you nowhere. You must have both. Therefore, input the information immediately. If you get a bill for concrete, don't send it to the field to verify its quantity and then to accounting to check the extension of the figures. Put it into the system immediately as an incurred cost. If it has to be changed later, you can change it in the system because having it entered is more important than finding out later you get another 2 percent discount or even that the invoice is wrong by 50 percent. There are usually many invoices missing from the system because we haven't received them. The fact that you haven't received an invoice doesn't mean you haven't incurred a cost. You need a system that will understate the profit rather than overstate it. You need this for a very simple reason. When profit is overstated, we take no action. We're pleased with the way things are going. But when it's understated, we will be less pleased and take steps to improve performance.

9.9 COST CONTROL VERSUS GENERAL LEDGER

Another thing to keep in mind is that the cost system doesn't need to tie into the other accounting systems in the company. It's fine if it does, but not if it's a slave to the other system; that is, the system has to wait for data until the other systems receive approvals for accuracy. If the latter is the case, use an independent cost control system and don't compare the numbers at all. It is far more important to get this critical management information on a timely basis. The other accounting information can be used for other purposes.

9.10 TRACKING COSTS

The process of accounting for all costs may be more complicated than it has to be if we track certain costs unnecessarily—for instance, subcontractor costs. Let's take the example of a building construction project where 60 percent of the project is subcontracted. In this case the costs we can actually control are the remaining 40 percent, and we should track those costs closely. Put another way, if you sign an electrical subcontract for a fixed sum, the money can be considered reserved, if not spent. Excluding change orders, the full amount of this electrical subcontract will be incurred as the job gets done. Knowing how much to pay against the contract as the job progresses is a totally different subject. These figures don't come from the cost control system. Quite the opposite. When the subcontractor is paid, that information is fed back into the control system. Therefore, it makes good sense to list, in a cost control system, the full contract amount as committed or spent as soon as the contract is signed. Of course, the amount is listed right next to the estimated amount. You don't need to wait until the end of the job to know that if your bid was $900,000 for electric and you awarded it for $850,000, you made $50,000 on the line item (or lost if you had to award it for more). The money is committed and gone. The profit (or loss) is incurred, so we don't want to mix it with the cost control figures we are tracking monthly to manage the work. The reason is simple. If our totals reflect the $50,000 we made on the electrical subcontract, the job will look better than it is by that amount. Theoretically, we could do $50,000 worse on other work before we realize that we're not meeting our estimates on the entire project. If each line item stands on its own and if subcontract and material prices are considered spent as awarded as far as the cost control system is concerned, the process is simplified and the tool is easier to use.

9.11 SUMMARY

If the cost control information you are now getting isn't accurate or isn't on time, stop using it. Bad information is worse than none. In these circumstances the alternative is much better—trust your gut feelings. If you spend any amount of time in the field, you probably have as good a feel as necessary to make a judgment on whether things are going well or not. Act on that judgment. If you don't get to the field, talk to the people who do and find out what their gut feelings are. The people you should talk to are those who visit the field periodically, not those stationed there full time. This is because the full-time field people usually have an optimistic view of how things are going and/or are hesitant to report bad news. If you don't get a good feeling about what you're hearing, go to the field and find out for yourself. It's time to use all of that experience and know-how that's part and parcel of being a contractor. What you sense on a field visit is important, and you need to follow your instincts even if the paperwork reveals the opposite. Is the site noisy enough? Is there enough excitement? How does it feel to you? Your gut responses will be right most of the time, and you should rely on them until you can establish an effective cost control system.

It seems almost silly to say that all contractors who fail do so because they didn't make money in the field. But it ceases to be funny when I tell you that more than half of the distressed contractors that I worked with didn't know they were failing until it was too late to do anything about it. The failure to evaluate contract profitability by project by month is one of the most serious and most avoidable causes of contractor failure. Cost control is fundamental to the business, and managing construction for a profit is your job. There is one thing worse than losing money on construction, losing money and not knowing it.

CHAPTER 10

Equipment Cost Control

10.1 OWNERSHIP COSTS

The use and ownership of equipment affects every type of contractor. The process of evaluating and accounting for all costs associated with equipment usage and ownership is complicated by the fact that some costs do not appear on invoices and, in fact, aren't incurred in the course of operations. These costs, while not easily recognized, will eventually have to be paid. Although various formulas are used to cope with equipment costs, the total costs of ownership are sometimes misunderstood. All costs, including those hidden, must be recognized and planned for in advance.

Contractors face many equipment concerns, like whether to buy or lease, which equipment to invest in, and when to invest. The equipment-intensive contractor has to make these decisions often, and they have a rapid and profound effect on his success. Other contractors have the same concerns, but less effort may be put into these decisions because equipment isn't a big part of their business, creating other problems. (See *Construction Equipment Guide*, Second Edition, by Day and Benjamin.)

Debt service and maintenance costs exist for contractors who own very little equipment and decide to buy a little more; these costs exist in a big way for equipment-intensive contractors. While we will be discussing primarily problems facing the equipment-intensive contractor here, the principles apply to all contractors who own equipment.

10.2 HOW MUCH TO OWN

The first step in controlling equipment costs is to control the amount of equipment owned. The decision to purchase new equipment is made for basically two rea-

These early examples of heavy equipment were powerful tools and the forerunners of the machinery that helps build our country. (Courtesy of Caterpillar, Inc.)

sons—to replace aging equipment or for expansion—and both are certainly valid business reasons. Nonetheless, both reasons must be considered judiciously because the contractor is usually committing a great deal of money without assurance of future work.

In replacing equipment, the contractor must weigh very heavily whether the new equipment is really more productive than what he has, and if so, by what margin? Is that difference worth the investment? He must ask if the existing piece of equipment is a maintenance headache or if he should invest in a complete reconditioning and get three or four more years out of it. Is his next two to five years of work a certainty? Is the marketplace growing or shrinking, or is it likely to change soon in either direction? Of course, the contractor can't answer those questions with certainty.

So the decision to purchase new equipment, which must produce profits over a period of years or be a financial liability, is not an easy or simple one. The decision to buy means taking on additional costs and creating a necessity to get at least enough work to keep the equipment busy. Too often, contractors *want* to buy newer and bigger equipment rather than need to buy it. When they need a replacement, some assume that bigger is better. Sometimes they even recondition the existing machinery to supplement the new machinery because the old equipment is more appropriate to their work.

10.3 REASONS TO BUY

The decision to buy additional equipment for expansion is usually made for one of two reasons: (a) new work is already contracted and there is no owned equipment available to do it or (b) the contractor is in an expanding marketplace and wants to have the equipment on hand to do the anticipated greater volume.

Let's look at the latter first. If your marketplace is getting stronger and is growing, it may be reasonable to assume that you will get your share of the growth and therefore greater volume. The problem with buying or committing to more equipment in advance of getting the work is, as already mentioned, you then *must* get more work just to keep the equipment busy.

10.4 COMPETITIVE POSITION

One of the difficulties in getting this new amount of work is that your competitive position relative to your marketplace may not stay the same when the marketplace grows. Your regular competition may also have a bigger appetite and may be going after the work more aggressively than they did in the past. Another situation that often develops when a marketplace gets stronger is the influx of outside competition. When out-of-area contractors are drawn to a strong or growing marketplace, they need to get a foothold and often bid very tight to get the first job or two. You may get less work and not more, and you will be forced to bid tighter. New equipment becomes a real burden when it forces you to go after a greater amount of work at a time when you must bid work at a lower markup in order to get it.

Even if you have contracted for more work than your current equipment will bear, the conditions of an expanding marketplace still make purchase risky. Once your existing work starts to finish up and the equipment used on those jobs becomes available, you may have idle machinery. Increasing your inventory of equipment should be very carefully thought out, not only as it relates to the work, but also as it relates to the marketplace, your competitive position, and your resources to do the additional work at a profit. I can't tell you how many successful contractors I have seen load up on equipment in good times only to be forced into severe difficulties by the very same equipment when the marketplace went back to normal. Their equipment began to run these contractors instead of the other way around.

10.5   CALCULATING EQUIPMENT COSTS

The subject of calculating and accounting for owned equipment cost is one that is ignored by some contractors or lost sight of by others who believe their accountants are taking care of it. To estimate work and bid a job, a contractor needs to know exactly how much equipment is going to cost him per unit time, and he has to include all maintenance and replacement costs to do that. He must have this information to know whether he's bidding at a profit, even if the market won't bear his profit margin, and to know whether individual jobs are profitable. The basic concept for costing of equipment is quite simple, but calculating it can be another story.

10.6   TIME AND USAGE

The cost of owning equipment is a function of both time and usage. Let's start with equipment that will be busy all the time under normal one-shift-per-day conditions, for example, a rock crusher. (The example works for 12-month or seasonal contractors.) The contractor owns it when it starts on a particular job. He bought it new, outright. Now the direct costs to him during the first month of operation, assuming mobilization is charged separately, are fuel, insurance, and regular maintenance. But there will be major maintenance as the project progresses, and spare parts will be required, so he must include in his equipment costs an allowance for this. The allowance is an estimated cost that will be tracked and corrected occasionally to reflect what really occurs. The allowance is treated as an incurred cost because in

As long as the equipment is working, most equipment accounting methods capture all the costs associated with ownership. A prudent contractor accounts for costs whether or not the equipment is active. (Courtesy of Caterpillar, Inc.)

three months, the crusher will need new belts and a bearing, and the cost for these is not chargeable to the third month. The belts and bearing were consumed over the three months, and their maintenance should be charged in a timely manner by estimating in an attempt to reflect this reality. These very real equipment costs are captured to account for the true cost of ownership, but because some are yet to be incurred, these costs are not applied to the general ledger.

Considering estimated maintenance costs as incurred monthly costs means including all maintenance, regular and extraordinary, to keep the equipment operable and in the condition it was when it came to the site less normal wear and tear (which we account for in another column). Because a contractor must know his monthly costs, the estimated cost is applied in his cost control the same as incurred costs and the estimate corrected periodically (usually annually). If the equipment will need major overhauls such as engine replacements after two or three years, these costs have to be factored into the estimated maintenance and accounted for from the first month the equipment is put into service. If this is not done, the contractor will be overstating his real profit by not charging the use or wear of the equipment to the jobs that caused it. His becomes a false economy.

10.7 REPLACEMENT COSTS

The contractor enables himself to replace machinery by charging replacement cost (not purchase price) to cost control monthly as an incurred cost. The replacement value is calculated by determining the useful life of the equipment and estimating the replacement cost less salvage at the end of that life. The contractor then divides the replacement cost by the useful life to get the monthly cost he will incur. He uses the number of months worked per year. This is the cost to him of using up the piece of equipment. If he takes only the purchase price and divides by the useful life, he won't accumulate the replacement cost of the equipment because of inflation.

One contractor asked me, "Do you really want me to charge my clients for next year's inflation when I'm only working for them this year?" I answered, "No. You can always pay for it yourself." And if you do, you will be paying for the privilege of being in business.

10.8 EQUIPMENT COSTS CHARGED TO PROJECTS

The purpose of charging all equipment costs to the jobs and applying these charges monthly as costs incurred in a cost control system is to give the contractor a realistic picture of whether he is making or losing money in time to do something about a losing situation. Equipment is not an overhead cost. Moving equipment back to the yard doesn't stop the ongoing cost.

Consider the heavy equipment contractor during a slow period. He can either charge more of his equipment costs to fewer jobs or eat the difference.

There will be no fuel cost, and maintenance can be suspended, but the insurance cost goes on as does replacement cost, which, while partially a function of usage, will be practically the same. It is obsolescence that really causes replacement so the timing of replacement won't change much according to usage. If a contractor believes that downtime will extend the useful life of the equipment, then he can adjust the replacement cost in the monthly figure as long as he factors in an amount for deterioration from storage and nonusage. Deterioration can be a costly factor because most construction equipment wears better in use than out of use.

10.9 IDLE EQUIPMENT

The alternatives open to a contractor whose equipment is idle because of an inability to contract profitable work are not encouraging. To take work on a tight schedule isn't good business because the contractor takes on too much risk just to keep the equipment working. To take break-even work is never justified except for survival. Liquidating some equipment is an alternative but must be considered in the context of the overall business, including new work anticipated. There is seldom a profit to be made in liquidating used construction equipment, although liquidation can reduce losses caused by the ongoing cost of idle equipment. Leasing out idle equipment is a favorable alternative, but this is usually difficult to do if the equipment is idled from a general slowdown in the market.

If nothing can be done to mitigate the loss from idle equipment, it should be left on the last job it worked on and the real costs of owning it charged monthly to the job. This serves as a reminder that the equipment is idle, and management of that job, who should have anticipated when the equipment would be free, will be encouraged to get everyone else talking about where it should go next. It also tends to discourage project people from always for more equipment than they really need.

If idle equipment makes one or more jobs show losses by the month, it simply points out the real costs being incurred. If all jobs are showing a profit but a lot of equipment is idle, the real picture for the company is not as good as the paperwork is showing.

10.10 CASH FLOW

Even if a contractor accepts these concepts and accounts for all of the costs as described, his cash flow will be greater than his real profit, and if the replacement and extraordinary maintenance money isn't put into reserves, then the money isn't there when it's needed. During slow periods with a lot of idle equipment a contractor could be showing losses on all jobs but still have a positive cash flow and so weathers the storm well. But if he doesn't make up the funds reserved for replacement out of future profits, there isn't going to be enough money to replace the equipment when the time comes.

Because most equipment is purchased not with cash but on credit, the equipment is expected to be paid off from future work. I used the example of equipment purchased for cash because it makes the concepts easier to relate to and follow. For equipment purchased on credit there are only two changes in the proposition. Interest costs are added to the formula as an expense similar to fuel and insurance. During idle time or when losses occur for any reason, instead of the cost control figures showing a loss concurrently with a positive cash flow, there may be a loss with a negative cash flow. If this happens, there may not be enough cash flow to make the equipment payments.

10.11 EQUIPMENT OBSOLESCENCE

Equipment-intensive contractors like heavy/highway contractors, landscapers, sheetmetal contractors, and others incur a great deal of cost in replacing equipment as it wears out. They have another exposure in the equipment area that is not as apparent and often not planned for, and that is equipment obsolescence. The productivity of their equipment dramatically affects the profitability of the equipment-intensive contractor and is part of his competitive edge. Equipment productivity is critical to making a profit and to bidding and getting the work. As newer and more productive equipment comes on the market and his competitors buy it, the contractor can be forced into equipment replacement earlier than planned just to remain competitive.

10.12 EQUIPMENT OBSOLESCENCE CASE STUDY

Consider the sheetmetal contractor with aging duct-making machinery who faced new and unexpected competition from a start-up contractor who had the latest computer-run equipment. The productivity of the automated equipment allowed the start-up contractor to bid lower on every job of any size that came out during the next year until the established contractor had hardly any work. He decided the only way to remain in business was to replace his equipment with the more productive machinery his competition was using. Like most contractors, he had not reserved money for equipment replacement or for that matter even accounted for it. In fact, the equipment was so expensive that his last five years' profit wouldn't have paid for it. His current financial statement reflected a bad year because of the new competition, and he was unable to secure the financing for the new equipment and subsequently failed. While equipment obsolescence was not a subject he considered very often, the contractor was quite familiar with the latest developments in his field and the capabilities of the new equipment that was continually coming onto the market. In fact, he knew that eventually he would need to replace some or all of his equipment, but for the time being he felt that it had a lot of good years left in it. He was not, however, including replacement costs anywhere in his cost accounting. By ignoring the real cost of replacing his equipment, he was enjoying a false profit. Had he accrued realistic replacement reserves, he would have seen that his real profits weren't what he thought they were, and had he considered obsolescence, he may have planned for continual upgrading or at least measured how far he was falling behind the state of the art.

10.13 REPLACEMENT COST IS INCURRED DAILY

The entire, adjusted-for-inflation, future replacement cost of equipment necessary to remain in business will become due whether or not we account for it. It is a very real cost of doing business, and it is a cost today, not at replacement time. Let's take a simplistic example. If I decide to go into the dirt-moving business and buy a $100,000 bulldozer, how shall I account for the ownership of this piece of equipment in years to come? Let's say I buy it for cash and that it will last for five years, at which time it will be worn out, and for our example we'll assume zero salvage

As newer and more productive equipment is developed, competitive pressure can acceler-
ate the need to reinvest in new equipment. (Courtesy of Park Construction Company and
Caterpillar, Inc.)

value. There are a number of ways to account for depreciation, and we'll select straight-line depreciation over five years or depreciation of $20,000 a year. If I recover all other usage costs during that five years and charge only $20,000 depreciation in my accounting, where will I be in five years? I will no longer have my $100,000 because I spent it to buy the bulldozer in the first place. I won't have the bulldozer because it is worn out and has no salvage value (use 8, 10, or 12 years if you like; it comes out the same), and I don't have a job because I don't have the piece of equipment. What happened to the $100,000? It's been consumed by the business. Sure I made profits during those years and the depreciation allowed me to have $20,000 of the profit without corporate tax, but the $100,000 was after-tax dollars. A new bulldozer at this point costs more than $100,000 (which it cost five years ago), say, $150,000, and to stay in business I need to borrow the $150,000. Now I not only do not have the $100,000 that I started with, but I am also in debt another $150,000.

The example is admittedly simplistic, and there are other ways to describe the same scenario, but there is food for thought here. For one thing, replacement costs will become due at some point and will almost always be more than the original purchase price, so we cannot rely on allowable depreciation alone to accurately account for the cost of ownership of our equipment. If we buy equipment on credit, as most contractors do, and replace it with credit purchases, we will go deeper in debt by at least the rate of inflation the longer we remain in business. While the IRS does not allow us to reserve money for replacement cost of equipment as a cost of doing business, that doesn't mean it isn't done. When we don't account for the real replacement cost of our equipment, it exaggerates our profit, which gives us a false picture of where we are and certainly of where we're going.

I get a lot of debate on this subject from accountants and tax experts, and I'm not lobbying to change generally accepted accounting practice or IRS regulations. I'm simply suggesting that to ignore the real replacement cost of equipment necessary to remain in business can allow us to operate in a false economy, go farther into debt over time, and for some contractors create serious long-term problems. Costs that are incurred and due in the current accounting period are no more real than costs that will definitely be incurred and will become due but not until a subsequent accounting period; it is just harder to recognize and account for them. Some of those costs won't be incurred for five or more years and are not a big problem unless you intend to be in business longer than that.

CHAPTER 11

Billing Procedures

11.1 GUARANTEES

The billing procedures in the construction industry are unique. The contractual arrangements guaranteeing and assuring the performance of the contractor and payment by the owner are one-sided, giving the owner more protection than the contractor. On public work and on a great deal of the larger private work, 100 percent payment and performance bonds are required. This effectively assures the owner that if the contractor is unable to finish the job at the agreed-upon price, the bonding company will pay the difference. However, the contractor on most projects has no assurance whatsoever that the owner even has the money or the decency to pay in a timely manner. We know of public institutions entering into contract and then being unable to pay for them. Most contractors doing federal projects know someone who performed signed change-order work and couldn't get paid because the owner or his representative didn't have the proper authorization.

11.2 RETAINAGE

Even with 100 percent surety guarantees the owners also can withhold part of every payment as a further guarantee that the contractor will do the job. Retainages of 10 percent are common, and only in recent years have some retainages been reduced to 5 percent when a job is 50 percent completed (and acceptable). Finally, the federal government has begun to deal with the inequities of huge retainages on their projects. Retainages are an expense to a contractor, which he, hopefully, passes on to the owner in his bid. But the assumption involved is that the contractor probably won't do his job unless he is forced. Owners deny this. The authors of the contracts deny it. But no one wants to give up the practice. Its very usage sets the tone for the relationship between owners and contractors.

11.3 ATTITUDES TOWARD CONTRACTORS

In the language of most contracts and in the role (supposedly independent) of most designers in the construction process, there is an assumption that the contractor isn't going to do the job he's agreed to do unless someone holds a gun to his head. The environment in which we work as contractors is one of mistrust on the part of the designers and owners. Retainage helps to perpetuate that environment. Trying to get retainage released at the end of a project from designers and owners who deny any built-in bias or mistrust further perpetuates the environment. When a million dollars of retainages is held for months awaiting $50,000 worth of punch-list work, it becomes obvious that mistrust exists and that the contractual arrangements provide financial advantages to the owner. Everyone knows that someone is getting interest on the $1 million retainage while contractors are paying interest on loans to compensate for high retainage.

11.4 ENTITLEMENT

This environment causes many contractors to feel they have very little power as far as their payments are concerned and that entitlement to their money is somehow clouded. Contractors want their money, and some really go after it, but often even they don't seem to have a strong feeling of entitlement. This condition is reinforced by the practice of walking a site with an architect or his representative at the end of a month to agree on the percentages of completion of the various line items of work. The architect's representative and the contractor's representative usually have a copy of the last month's payment requisition in front of them. They agree, or debate and then agree, on what was done that month.

This practice is common on unit price jobs but is also very prevalent on lump-sum projects. Some contractors don't do this. They send their completed requisition to the architect. If the architect disagrees, he sends it back marked with his changes (usually in red pencil), and the contractor retypes it, signs it, and sends it back to the architect. These practices demonstrate the payment environment in the industry and demonstrate, for me at least, a problem with entitlement. These practices put contractors in the passive role of taking what they can get rather than the active role of invoicing their customers for services rendered and product delivered.

The very word "requisition" suggests we can only ask for our money. Furthermore, it suggests that there is some question as to our entitlement to it. I have had contractors tell me time and again that this cooperative approach is the smoothest and most expedient way to get the requisition approved and passed on to the owner for payment. They usually add, "Besides, the architect always treats me fairly." I suggest that in the name of fairness we have tolerated for too long the attitude that we need to be controlled and that the designers have to protect the owners from us. I think that we are *entitled* to "fairness" from all parties to the contract. Fairness is a two-way street. It is not something we have to ask for or earn. We don't get it, and it's our own fault for not demanding that to which we're entitled.

11.5 WHO PREPARES THE REQUEST?

According to most contracts, the payment requisition is to be filled out by the contractor and approved by the designer. I, for one, don't need any help determining what work was performed during a month nor how much I want to be paid for it. If the designer doesn't want to approve it, that's his prerogative. He can red pencil it and send it on to the owner along with the amounts he thinks should be paid. There's no need to retype it, nor is there a need to walk the site and bargain for the amounts. All you establish by doing those things is that he (the designer), and only he, decides what you will be paid. Retyping means changing your request to say what he says you have to say. What we do by these practices is place our monthly payment amount totally in the hands of the designer without even a *right of appeal*. Sure we can argue and cajole. But once we have agreed that only a clean, typed requisition will go to the owner, what can we really do?

Effectively what we have done is establish that we and the designer will agree on the amount before it is "approved" by him. This is ridiculous. It not only rewrites the payment provision of the contract, it also leaves us in the one-sided position of trying to talk someone into agreeing on how much we should be paid. But the designers don't have to pay the suppliers. They haven't already paid for labor. They aren't under any time constraints. So why do some contractors get involved in these practices? Some have told me, "It's easier than arguing over payment, because arguing will only cause other problems with the designer." Others have said, "I don't want the owner to see a red penciled requisition, because he'll think I was asking for more than I should have." And still others say, "The owner's not going to pay any more than the architect approves anyway so what's the difference?"

11.6 CONTROL

The difference is that the practice gives the designers one more element of control over us during the construction process. But as long as we intend to live up to our end of the contract (including every item in the plans and specifications, as most designers insist we do anyway), we have nothing to fear from the designer. We want his cooperation, but we do not need to earn it or pay for it in any way because we are entitled to it under the terms of the contract and as professions. There is no need to shelter the owner from our payment requests. He should have an active role in the construction process. One of the few things the contract says he has to do is to pay, and I like involving him in the payment process early on in a project.

11.7 COOPERATION

If this sounds like I'm not in favor of cooperation, that's wrong. I'm just in favor of cooperation on sensible and businesslike terms. If the cooperation is one-sided to the point that the contractor keeps quiet so the owner is always happy and the designer's boat isn't rocked, that's nonsense. But it happens. And it starts with the payment process when a contractor doesn't stand up for his rights under one of the few sections of the contract that are favorable to him. He is supposed to requisition for work performed, and the owner is to pay within a specified period of time. If the designer doesn't approve the full request, the owner only has to pay the approved amount. But contracts don't stipulate the consequences of designer disapproval. So the designer appears to hold all the cards. The contractor is simply doing what the contract calls for, as is the owner, in paying only the approved amount. The designer, on the other hand, has stuck his neck out. If there are any consequences to the architect's actions, the owner will end up paying for them. And don't think the owner doesn't know that. Basically, we've tricked ourselves into believing that we are powerless.

11.8 EXAMPLE

If you don't think the practice is widespread, you're wrong. I once mailed back to an architectural firm a red-penciled requisition they had sent to me for retyping, with a note advising them to send it on to the owner as it was. I got a call from an incredulous architect saying that the firm had never sent an owner anything but a clean typed requisition. It was my first requisition on the project, and I always made the first requisition exactly or slightly less than the quantities completed so that I would be accurate and prepared to meet just this type of problem.

The architect had reduced most of the quantities, probably out of habit or perhaps because he thought he should. But the amounts were already understated, and one line item requested nothing where anyone could see that the work was obviously completed. At any rate, he said that he wouldn't send the red penciled copy to the owner, and I would just have to retype it and sign a new one. I told him that I never retyped a requisition but that I had sent the owner a copy of his penciled copy along with a copy of his letter telling me to change it and that I was waiting to see what the owner wanted to do.

The architect then proceeded to inform me that the contract stated that I could have no contact with the owner except through the architect. I told him that he would have to explain that to the owner. He said, "What if I send the owner the copy of the marked-up requisition and tell him not to pay any of it?" Both he and I knew he would have a hard time explaining that, and it sounded too much like discipline for me to even answer. The owner was advised not to pay. When the time ran out according to the contract, I gave a ten-day notice of default for nonpayment. That brought everything to a head and to a meeting with the owner at which the architect said things like: "Well, I don't know exactly what was done by the last day of last month," which pretty well clarified the situation for the owner. I was paid in full immediately and before the tenth of every month as specified for the rest of the project. Was the inspection on the job any more difficult than usual? I'm not sure. But I was sure that I wanted to be paid the correct amounts on time more than I wanted an architect who would overlook any poor workmanship he might find.

Slow pay and retainage held too long are an increasing problem in our industry. Too many people are allowing it to get worse by saying, "We can't do anything about it." Each contractor has to decide for himself what his approach to this subject should be after considering the impact of slow pay on his business. If we could calculate all the interest paid on lines of credit required because of overheld retainage, we would be astounded. And if we had the interest we could earn if we were paid for our work on time, our industry wouldn't be so high risk.

11.9 BILL ON TIME

One of the first things a contractor can do to collect his money faster, if not on time, is to prepare and send out his requisitions on time. I see bills go out on the 5th and 6th, even the 10th of the month when the contract specifically states: Bill for work completed as of the last day of the month and to be paid no later than the 15th (20th, 25th, or whatever) of the following month. You can't expect owners and designers to begin the payment process until you get the payment application to them.

Begin the billing preparation well before the last day of each month. Have a tentative requisition done by the 26th or 27th, anticipating work that will be completed by the last day, and verify it by phone. In that way the requisition can be sent by the fastest method on the 1st. Phone to make sure the bill arrived and isn't sitting on someone's desk. It's best to let people know that prompt and proper payment is important to you. They'll respect that. It's the sign of a good businessman. It's hard to change attitudes midjob after everyone's gotten used to a way of doing things. So you need to start from the first requisition on each job and be consistent.

11.10 CLARIFYING THE PROCEDURE

There is nothing wrong with expecting to be paid for your work and asking for your money. Get it straight at the initial meeting. After everyone tells you what they expect or need, tell them what you need. Simply be upfront and say, "I want to talk about the payment process." It's not going to be important to anyone unless you make it important, unless you let everyone know you're not embarrassed about it and you expect it to be dealt with up front and handled on time. Payment is not a backroom discussion. You can say that your efficiency and productivity depends on paying your subcontractors and suppliers on time and that you don't want to invest in the job—just build it. They know that and shouldn't need to be told, but they do. Remember, manage the process and keep it in the forefront of everyone's mind. No one else will do it for you.

After clarifying the payment procedures before the first requisition is due and making your needs and expectations known, be accurate. Submit a requisition for exact quantities of work completed. This way, if there is a problem with the designer, you can move with confidence right from the start, knowing he will be proved wrong if he red-lines your requisition. They'll get accustomed to taking your word for work completed. You'll get a reputation for fair rather than inflated requisitions. It's a fine strategy.

11.11 DUE DATE

Almost all payment clauses say that payment is to be made by the 15th, 20th, or such, but many owners take the attitude that it's not due *until* the 15th or 20th. The immediate consequence of thinking payment isn't due until a certain date is to assume that it isn't overdue until some kind of grace period, say, ten days. There are a lot of owners who feel that way. *By the 20th* means just that. It will only continue to mean just that in the construction industry if we insist on it. I talked to a public official once about getting payment for a job. He told me it was *approved* but that he did not have the authority to pay before it was *due*. We got the contract and reviewed the payment section, which said *by the 20th of each month*. He said, "Then what is the due date?" I replied, "The payment is due upon approval of the architect, but no later than the 20th of the month." Without another word he authorized the check. I'm not suggesting that getting paid is easy by any means, but I am suggesting that we need to be aggressive in getting our money. Furthermore, we should read our payment provisions and explain them to the owners.

11.12 HIDDEN COSTS OF RETAINAGE

The cost of accepting late payments is obvious, but the cost of late retainages may exceed even the interest we must pay for the wait. When a contractor gets stretched out and needs his retainage and when the amount of retainage far outweighs the value of the missing items, a lot of contractors give away the store to get paid. They either do work that they don't owe the owner or cut a deal forgiving part of the retainage in return for collecting the rest. The forgiven amounts are ostensibly for work not done, but they are usually pure concessions. This happens often on jobs where a contractor *cooperates* throughout the project. Consequently, I strongly recommend that we stand firm on all payment requirements from the first requisition. If we don't get treated in a businesslike manner when it comes to getting paid, it may be because we haven't acted businesslike and therefore can't *demand* businesslike treatment.

11.13 SUMMARY

Not getting paid in a timely manner takes some of the fun out of the construction business. It also adds to the risk. Cash flow is always a concern, but when a contractor's marketplace weakens and his work slows down, cash flow problems can become critical very quickly. If all of a contractor's reserves are tied up in his receivables, he can be forced out of business. If many of the receivables are old retainages and slow payments, the contractor has to shoulder part of the blame for not demanding what is rightfully his. If the owner doesn't fulfill his side, don't take on his problems. A completed building is excellent collateral. Don't go under because someone is failing to honor a contract that you've completed.

CHAPTER 12

The Use and Misuse of Computers

12.1 A TOOL

The computer is perhaps the best piece of construction equipment ever added to the industry's inventory. It has provided an efficient way of dealing with a tremendous amount of details. It has revolutionized the scheduling process, given us entirely new systems for estimating, and provided excellent accounting data storage and retrieval. Computers have also caused problems. The transition from manual to electronic systems and from one electronic system to another is often a very trying and costly experience. Determining the size and kind of equipment and software you need is also difficult. The best attitude to take is one that sees the computer as a tool. It will be an enormous help if you get the right-sized tool, learn to use it correctly, maintain it, and respect the fact that if mishandled, it can be dangerous to your company's health.

12.2 DATA LOSS

Throughout this book we have been discussing the need for accurate and timely data to manage a construction business and the risks inherent in trying to manage without good data. During the conversion from one computer system to a larger one for expansion or to a different one for increased sophistication or from an outside service to an in-house service, you risk a temporary or even permanent loss of data. The conversion from a manual system (which some smaller contractors still use) to a computer carries the same risk. Most of the transitions I've seen are scheduled to take place between monthly reporting periods but often fall short of that goal. It is

Computers have developed dramatically over the years from big pieces of equipment used by government and larger corporations to much smaller, very powerful units practical for almost any size business. (Courtesy of IBM.)

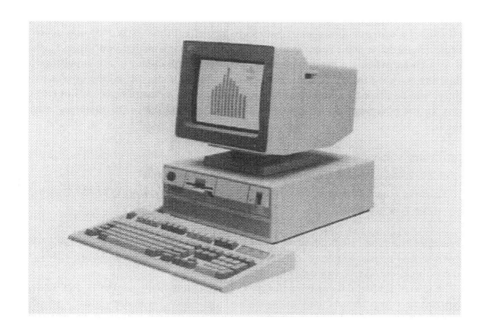

not uncommon to find errors in the data transfer that render all the reports wrong at first. It often takes months to get the situation straightened out. In the meantime, the contractor has no management information. Fortunately, in most of these cases, the errors were so huge that it was obvious that the reports were wrong. When the errors are small, they can go undetected and the erroneous data accepted as correct.

12.3 COMPUTER DEPENDENCE

The preceding problems by themselves do not threaten a company's survival, but they can be expensive and a serious disruption. Computer dependence, however, is one of the risks you will have to face. When a computer goes down, it isn't easy or sometimes even possible to return to manual systems. I've observed companies unable to do their billing for two months. They had to negotiate with customers to accept estimated bills after a conversion mistake. I've seen companies unable to process payrolls for weeks. I've even seen one company who had its bonding suspended when it couldn't produce a financial statement because of a conversion problem.

12.4 CONTROLLING THE RISK

There is business risk in our computer dependence. It should be dealt with by establishing a formal contingency plan to go into effect if the data or system is lost. Professional advice and even insurance is available to cover the contingency. The risk is greatest when we are intentionally transferring the data to a new system or tampering with it to upgrade the existing one. This risk should not be approached lightly, and contractors should be looking for assurances that the risk is limited by checking the planning and the process. The do's and don'ts in the planning stage include

do worry about it

don't take any vendor's word that it's a simple procedure

do have a backup copy of your data and a means to return to the old system if there is a problem

Sometimes the latter isn't possible, but by storing duplicate data off the premises (which should be part of your everyday loss prevention activity), you may be able to use an outside service in an emergency.

The loss of critical data or the inability to get accurate management reports while upgrading equipment or changing software will have a serious impact on any construction organization. For a contractor in difficult times who may be struggling already, it could be disastrous. Basically, understand that there is a risk, and the best way to overcome the risk is to run the two systems simultaneously until the new system proves itself. The costs are negligible compared to the cost of a system crash or partial data loss.

12.5 INFORMATION EXPLOSION

Some contractors suffer from not enough information from their accounting systems to properly manage their businesses. An increasing number of contractors today are suffering from too much information. We are undergoing an information explosion in the construction industry fueled by the advent of faster and cheaper computers, aggressive sales presentations, and the introduction of a new discipline to the contractor's staff—computer professionals. This combination has made more and better data available to construction people to manage their work; in some cases too much data has been made available. Because of all this data, we are forced to decide who within our organization should get what information, and some people are taking what seems at first like the best approach and giving everyone everything.

Once information is collected, using it in many ways and producing numerous reports is easy and printing out extra copies is cheap, but we can also get buried in all this paper. In some cases we are sending out so much information that no one has time to read it. Worse, if we are sending too much information to the field, either it is ignored because of its bulk or it is so distracting that people don't have time left to supervise the work. From an industry that 20 years ago barely sent labor reports to the field, we are sometimes sending 2, 3, and 4 inches of computer printouts a week. That the information is available does not mean we need it all. Those generating the reports tend to want to fill an entire sheet of computer paper or to add columns that aren't necessary or important. I've seen reports that string out for each vendor original contract amount, paid to date, percentage paid to date, unpaid to date, percentage unpaid to date, and so on, and so on. Sending information like this to people who have no use for it is data overload and makes one wade through 2 inches of paper looking for information that is not needed or just give up and stack the report with all the others in case someone asks for it.

12.6 FIELD PROBLEMS

When unnecessary information is being sent to the field, a more serious problem exists, that is, diminishing respect for the administration side of their organization. I have field people saying to me, "If this is the kind of information they (the office) think I need, they don't know what they're doing."

The field is where we make our money, where we earn our living, and the administrative and accounting side of the business is to service the field force. It is up to us to find out what they want in the way of information and deliver it, not decide for them what we think they need and bury them in it.

12.7 FIELD TEST

In my consulting work when I thought information overload was one of the problems a contractor was facing, I would recommend a particular test. The projects would be sent the exact same reports for three months. No date changes or alterations of any kind, simply the same monthly report that they had received one, two, then three months ago. In most instances there would be no complaint, which simply meant that no one was reading them. Some contractors would get upset at their field people over this, but I consider it management's problem. In some of these cases management needed to stop talking to and start listening to their field people. They had drifted apart, and management was busy putting in all kinds of new systems and sending all results to the field instead of telling the field people, "We now have this, this, and this information available. Which, if any of it, can you put to use?" or at the very least, "We'd like you to use this information in this way, what do you think?"

Information is a tool, and the user of the tool has to want it to learn how to use it. If we deliver a trim-saw to a concrete pour, it will just sit there next to the inappropriate computer reports. The contractor or other senior construction people need to be involved in the decision of what reports will be generated, what they will contain, and who will get them. We need to answer the question, "Will this information help the person receiving it do his job better and does the user agree?"

12.8 SUMMARY

There is the temptation to overuse any tool. In the construction industry, the computer will continue to have impact and will improve methods and procedures that were formerly done manually. As new things come out, there will be some mistakes made through inappropriate usage or because some new products just aren't perfected before they are used. Few of us want to be the first to try an unproven innovation, but few wish to be last. Care in selection and use of computer systems and automated procedures will be necessary as the industry continues to modernize.

CHAPTER 13

Other Industry Concerns

13.1 GROWTH AND RISK

The words ''growth'' and ''growing'' recur in the study of the management of risk in the construction business because the risks are simply greater during growth phases. A construction business must be managed well to be successful, and in the best of times there is risk. A rapidly expanding construction company magnifies its risks unless it is closely and intensely managed. In a volume-driven industry that thrives on growth there are failures even among the older and established firms. There is nothing wrong with building a bigger and bigger business. That's the American dream. But the increase in risk in the construction industry from growth alone cannot be understated and should not be overlooked.

13.2 MARKET DRIVEN

The ideal construction company would be organized to be market driven and not volume driven. It would strive for carefully planned growth but be prepared to level off or fall back on volume if the marketplace tightens or shrinks. It would use its markup flexibly as a competitive tool but never take break-even work just to maintain volume. In a tightening market (greater competition for the same work) or in a shrinking market (less work available), the ideal company would bid more competitively than it would in a better market but concentrate on making more profit on less work. It would have some flexible overhead built into the organization that could be cut immediately and would not hesitate to cut permanent overhead when necessary. Be willing to get small again to survive. The down cycle will pass and you'll be ready for the upswing only if you come through intact.

13.3 CONTROLLING THE NEED FOR VOLUME

Overhead costs are difficult enough for a contractor to control when he's not growing, but in a growing organization they pose two dangers. Because we cannot add a half-person or a half-piece of equipment, we are forced to put on overhead costs during growth in larger amounts than we'd like. This can cause losses until we grow into our overhead. Herein lies the two problems: Losing money for any length of time is dangerous, and needing additional volume as an absolute necessity to cover overhead puts us in jeopardy. Whenever we are demanding a greater market share, we can expect price to suffer. While we may not make a conscious decision to lower our price when we must have added volume, that is what usually happens. And when price suffers, it is usually for all our new work, not just part of it, so we end up needing even more volume than originally planned because our margin is suffering. This can lead to a downhill profit spiral during rapid growth.

The additional growth then requires more overhead, creating temporary losses and the immediate need for even more volume. This spiral has caused numerous construction business failures.

Rapid growth will also put a strain on the company's key people and systems, and substained growth doesn't allow for a reasonable training period. Of even greater concern, continued growth doesn't give a contractor a chance to test new people or systems before the next new people are put on and systems added to. If things aren't going well at any point, it's discovered after additional volume and people are taken on, and corrective measures are more difficult with everyone already stretched out and coping with the largest volume the company has ever handled. Again some companies don't recover from this scenario. Some cases just look like continuous growth was demanded with no measurement of performance right up to failure.

13.4 RATE OF GROWTH

If your market is not growing, growth is more difficult. Growth at a rate of more than 15 percent should be considered substantial, and sustained growth at a percentage of the previous year's volume compounds quickly. At 15 percent you double in 5 years and triple in 7, at 25 percent you double in 3 years and triple in 5, and at 50 percent you double in 20 months and are five times larger in 4 years. Growth of two and three times the previous size requires more resources in the way of people, systems, and money, and the larger company is untested as an organizational unit until it has operated for at least a year profitably and smoothly. If this test proves unsatisfactory but new growth has already been added, then you're looking at 2 bad years instead of 1 before you can roll back to your proven team. For some it never happens.

Incremental growth instead of sustained growth may seem unnatural, even unnecessary, but it is the best way to control the risk. With a series of growth, then test, then growth, then test again, you are able to recover after a bad test in lieu of constant growth until a bad year from which you may or may not recover. In sustained rapid growth you can grow out of your people and systems so often that you never really have the same organization long enough to truly test it, and you're at constant risk with a changing team. In some cases it's just a matter of time.

13.5 FLEXIBLE OVERHEAD

Flexible overhead is a new concept for the construction industry. Our marketplace is so unpredictable and affected by so many variables that it is difficult to forecast for even a few years. If we cannot be sure of a stable or growing market while we are causing or allowing our businesses to grow, we can control risk by putting on overhead to deal with our growth that can easily be removed if our market turns sour. With some of our overhead flexible we aren't slaves to our volume, and we can fall back and concentrate on profit. The method is to use temporary employee services for some clerical administrative and accounting functions. Use short-term rentals for some office and field equipment and short-term office leases, even temporary trailers, during growth stages until a new plateau of volume can be reasonably assured. Even management people can be brought on with specific company growth and performance goals associated with their continued employment. This creates challenges for new people, refocuses the real risks associated with growth for existing management, and has been done successfully by many start-up firms in this and other industries and more and more with growing organizations.

There are costs associated with flexible overhead as lease and rentals may cost more than purchased equipment, temporary employees may cost more, and efficiency may suffer. But the reduction and control of risk is well worth it, and the motivation and level of involvement of existing management people who get involved and excited about this cautious, realistic, businesslike approach to growth may more than make up for it.

Permanent overhead in a fickle market is dangerous. Flexible overhead may create cramped quarters and less creature comforts (privacy, plush offices, and the latest telephone systems), but those who use it to control risk during incremental growth phases say they sleep a lot better when they get home at night. Most who have tried the flexible overhead approach have put permanent overhead on even slower than would normally be considered safe and are committed to keep some portion of their overhead flexible at all times as a hedge against a market slump. They manage their profit and not their volume.

An excellent construction company would be as ready to do 25 percent less volume in any given year as it is to do 25 percent more and have no increase in risk either way. I realize this is a departure from the accepted norm, but I believe it is the profile of the successful contractor of the future.

13.6   THE PEAKS AND VALLEYS

Not that many years ago contractors stayed in their own backyards. They generally worked a lost closer to home because their equipment wasn't as mobile as it is today, short-term leasing wasn't as prevalent, and travel and relocation weren't as easy. There were always peaks and valleys in the marketplace like there are today, and when things were bad in a contractor's normal work area, he had to stick it out and do the best he could. But when things were good, he and all his competitors had a seller's market. Because contractors weren't that mobile, they didn't come in to a new area in great numbers and impact the market, so there remained the opportunity for substantial profits during good times. The expression "They took the good with the bad" is appropriate here. The good years allowed for great earnings, and in a more conservative era some of this "excess" would be put away as reserves against the lean years. Reserved or not, when a seller's market developed, contractors were able to generate substantially greater percentages of profit than we can today under any market conditions. The major reason is competition, which is a result of increased mobility. Today whenever a good market develops anywhere in the country, out-of-area contractors compete for a portion of it, invariably preventing a seller's market from developing and in many cases actually driving down prices. There are very mobile nationwide contractors that follow the good markets, and they are supplemented by contractors from any area of the country where there isn't enough work and who are willing to travel.

13.7   DIMINISHING PROFITS

The net effect is that peaks are taken out of our markets while the valleys remain. The opportunity for really big years is substantially reduced, and the average profit in the industry has diminished and shows every sign of staying down. Ease of mobility nationwide and internationally will continue or increase competitive pressures, which in turn keeps prices down.

What this means to the average contractor is that without the prospect of boom years to make up for bad ones, he must take care to control his valleys. With typically limited cash reserves a contractor can ill afford to increase risks with no controls. He must manage his businesses cautiously, if not defensively. Grow with prudence, test as you go, and be prepared to withdraw from bad decisions.

13.8   EMPLOYEE BENEFITS AND COMPENSATION

The subject of employee benefits and management perks ties in well following the concepts of *flexible overhead* and *peaks and valleys*. The general and administrative costs of doing business are as necessary to the running of a construction company as are the costs of concrete and steel. Controlling these costs is imperative. Overhead costs deriving from benefits and perks must be treated judiciously, and the best way to do that is to be miserly.

The discussion of bonuses naturally follows the management of overhead costs. Performance bonuses are quite common in the construction industry, and if they are part of a carefully considered compensation plan, which is known and understood by all of the participants, they can be very effective. Random, unorganized, and separate deal bonuses cause more problems than they are worth. Some companies have even fallen into the trap of giving bonuses each year regardless of company performance. Remember, they tend to become regarded as part of the wages and salaries unless they are tied to the performance of the employee or the company. These add overhead costs almost by accident, and the benefits diminish very rapidly. To be effective, bonuses must be part of a formal, overall compensation plan. They must be tied to each individual employee's performance, the profitability of the entire job, and the success of the entire company.

The cost of bonuses or unrealistic compensation packages put in during good years has accelerated the decline of many companies when lean years hit. Luxury automobiles, club memberships, and pleasure trips are near and dear to anyone's heart and commonplace perks for hard-working managers in many construction enterprises. The biggest problem with these overhead expenses is that the costs to maintain them keeps going up and up while benefit from them goes down as they become expected or are taken for granted. A company car is a tremendously valuable perk, and often it is given in lieu of a $2000 or $3000 raise in a particular year even though it's worth much more than that. The problem is that two or three years later it is sometimes taken for granted by the valued employee who now only understands one thing—that he is underpaid by $2000 or $3000 compared with somebody else. Giving such perks is hard to avoid because so many people are doing it, but there is greater value in having the highest paid people around with no perks than the lowest paid with great perks. You'll keep your employees longer and have no trouble getting new ones to quit their lower paid jobs to come to work for you because in most cases they will have lost sight of the real value of their perks. It's cheaper in the long run.

13.9   KEEPING GOOD HELP

Planning to keep employees for long periods of time in a family or clublike atmosphere is fairly common among small and medium-sized construction companies. It's also fairly expensive, and in the current working environment in this country it is becoming more difficult. With the mobility and lifestyle of middle management, two-career families, and a shrinking work force, people are changing jobs with greater frequency than ever before. Job security and company loyalty aren't the top concerns of today's work force. They are being replaced with quality-of-life issues and job satisfaction. Changing jobs or relocating without a job is becoming more common than remaining 15 or 20 years with the same company. Look back over your company history and recall who the key players were 5 or 7 years ago. For many this is an ever-changing scene, and it may be more so in the future. A close-knit group working in a clublike atmosphere can be very efficient, but if the players are ever changing, then a portion of the money spent on creating the family atmosphere might better be paid out in training replacements for key people and reserved for recruitment. Increasingly, the efficiently managed construction company is businesslike and professional, has a certain amount of internal competition among managers, and attempts to make its long-range plans around key positions, not key people.

13.10 INTERNAL COMPANY DISPUTES

The construction industry may be the largest industry in the country, but it is made up for the most part of hundreds of thousands of small and medium-sized businesses most of which are closely held or family businesses. Internal disputes among the management of closely held construction companies has created discomfort and disruption for a number of contractors and their families. In a high-risk, low-margin industry where businesses are often run at a high level of intensity and energy, some conflict can be expected. Some people seem to expect more from family members and are more tolerant of nonrelatives. For whatever reason, personal problems within a small or medium-sized, closely held construction company seem to cause more problems than they should and have been seen to affect company performance and profits significantly. Further, the succession of leadership in these types of companies is not as smooth as it should be, in some cases even when a great deal of planning has gone into it. It is an area of concern for owners of closely held businesses who need to be aware of it and provide for open and honest communication of all parties involved at all levels within and outside the organization.

13.11 CLAIMS

In all the ways that the construction industry has changed over the past 20 years the most onerous, frustrating, and costly is that of dispute settlement. There were disputes in years gone by, although certainly not in the number and severity we see today, but the resolution of disputes was for the most part more equitable and businesslike. In today's contracting environment with all parties in the building process trying to relieve themselves of any and all liability, we are left with too few clearly defined roles. Years ago construction professionals understood their areas of authority and responsibility without the need for arbitrators or judges; this has been all but lost to us. Contract documents get bigger, claims seminars get larger, and an entirely new group of services is offered to our industry—construction attorneys, claims consultants, and dispute avoidance specialists.

With the amount of defensive activities, paperwork, and energy that I see in many construction organizations, there is some question as to whether there is any time left over to devote to running the work and making a profit. The time that senior management spends today on legal issues and the expense associated with it is increasing at a frightening rate.

I am genuinely concerned that a new clause may eventually be added to the ten common causes of contractor failures, and that would be *dispute disasters,* companies that either couldn't afford to fight any longer or were simply blown away by the other side. Many will say you can't afford to do business without your lawyer, but you may eventually not be able to afford to do construction projects with your lawyer either. We need a truce to be called in the construction industry among the parties in the construction process. Not among their lawyers but among the parties themselves—the owners, the designers, and the contractors. Each has very specific responsibility and, yes, very specific liabilities. Attempting to avoid their respective liabilities and responsibilities causes most of the problems in this area, and a potential solution would be to go back to accepting them. I don't expect to see it happen, but if the party who makes the mistake fixes it at his own expense, you don't have a dispute. You would still have the problem and the cost of fixing it; you just don't have the cost of arguing about it. This, of course, is a little too idealistic for today's business climate.

Contractors need to recognize the risk of disputes and try to control them and the costs associated with them. The potential for disputes can increase with changes in project size and when working in unfamiliar areas or with unfamiliar owners and designers. Adding this exposure to the others discussed in this book suggests strongly that business expansion and growth be looked at carefully and planned prudently by contractors of every size and type.

When disputes do arise, they should be responded to quickly, and if the fault lies with you, it's cheaper to fix the fault rather than fight. When you're not at fault, be sure the solution doesn't cost more than the problem. Try dealing directly and fairly with the parties involved before expanding the dispute. If you're forced to litigate or arbitrate, try to limit the dispute to the original issue or issues and claim only your real costs. The theory of throwing in everything but the kitchen sink as long as you're in the dispute anyway just clouds the issues, complicates the process, and increases your cost of resolution. The idea of doubling everything because "they'll only cut it in half anyway" has backfired on a lot of people. What it tends to do is double the cost of resolution because it takes twice as long to weed out the padding and get back to the real numbers.

There are other exposures in disputes, and their resolution is distraction from the business, impact on the attitude and morale of the people in the organization, and the outside chance that an off-the-wall verdict could break a company.

13.12 CONTROLLED BORROWING

The construction industry uses credit in many ways. A contractor needs a line of credit with his surety company to secure a payment and performance bond on jobs that require them. He uses secured loans to purchase the equipment he uses to do the work. He may use bank credit to fund working capital either on a seasonal or as-needed basis or to fund growth. Bank credit is part of the everyday needs of a construction contractor, and the management of that credit requires skill and attention. Borrowing from a bank is not an event; it's a process. The process is another of the many things a contractor needs to manage if he is to remain successful. As a contractor-manager you must become concerned with the unplanned and unscheduled use of credit. It is too common for contractors to borrow working capital unexpectedly and not fully understand why the need arose. Be aggressive in determining why the money is needed today when there was no anticipation of the potential need last month.

If you find yourself borrowing working capital unexpectedly, it should signal the need for better cash flow planning or tell you that field performance is falling off. The construction business operating without some type of cash flow planning is out of control for the simple reason that you never know when you are going to run out of money.

A good line of credit is no substitute for cash flow planning. A company that doesn't borrow at all must still do cash flow planning. But for a company that borrows some or all of its working capital, cash flow planning is imperative. Not only is there the possibility of running out of both cash and credit, there is the added interest cost to be considered. I can guarantee that without cash flow planning, you are spending more on interest than you would with cash flow planning even if the planning isn't done well.

Cash flow planning is partially a state of mind and should be included in all decision-making processes. The first consideration in business is profit. The second consideration is cash flow: Will a change or a project create a cash outlay? Will it provide a cash influx? How soon? At what risk?

Borrowing should never be delegated. It is important to the security of the business, and is intricately involved in the process of controlling the risks the company faces at any given point in time. Borrowing should be controlled by top management through careful planning that takes into account the amounts and timing of credit use and addresses the sources and timing of payback. The need for any unplanned borrowing should be cause for great concern because either the cash flow plan is wrong or it isn't working. In either case, new planning is required. New plans on short notice should also be undertaken with the same diligence as the original effort.

13.13 BUSINESS PLANNING

Long-range and strategic planning are not addressed in any formal way by many contractors. That's not to say they can't tell you their objectives and the plans they have to achieve those objectives. What it means is that they don't have a written guide. They know their objectives but have no detailed plans to achieve them. Without at least an informal plan with some detail a businessman must react to whatever comes his way. It becomes very difficult to retain control, set the direction, and measure the progress in that direction.

Short- and long-range plans are the tracks on which your company runs. They make managing a construction business so much easier. It is surprising that contractors don't devote more time to developing and following strategic and long-range plans because doing so is such an effective tool. The time you spend in the planning phase comes back to you with incredible interest and dividends of time saved.

Planning should be done at a time set aside for just that purpose and outside the mainstream of daily activity. The owners and key managers of the firm should discuss and evaluate their individual and corporate goals and see how they fit. Not everyone wants to go to the same place, perhaps not even in the same direction. All the company's resources should be realistically evaluated and measured against short- and long-range goals to see if they fit. By establishing clear goals and directions that are understood by everyone concerned, meeting these objectives becomes easier if only because everyone is thinking along the same lines and looking in the same direction.

Things certainly don't always go as planned, but much of your business future is actually within your control. When things change, plans can be reevaluated and altered. This way, you're not simply reacting. You're acting in an organized fashion. The plan provides a measure of movement. I can't overstate the importance of formal, written, short- and long-range, detailed plans. I recommend one year of hard, or detailed, plans and three years of soft, or flexible, plans for medium-sized or larger companies and one year of hard plans and one year of soft plans for smaller or newer companies. They are the best risk control tools available. If you start the planning process, you'll find that you go about the job of managing with less frenzy and more confidence and purpose.

13.14 RECOMMENDATIONS

Be careful, prudent, businesslike, and professional as you manage your construction business; treat your employees, associates, and other parties in the construction process as you would like to be treated. Develop a business that you can truly manage with confidence so that some of the enjoyment comes back into building.

Glossary

Accepted bid. Bid that owner accepts as basis for entering into a contract with contractor who submitted the bid.

Accounting cycle. Complete sequence of procedures repeated, in same order, during each accounting period; e.g., recording and posting transactions.

Accounting principles. Foundations of accounting, based on accounting principles, from which practical rules arise.

Accounts receivable. Amount customer owes a business on a current account.

Accrual basis of accounting. Method of accounting incurred expenses and earned income for a defined period, regardless of whether expenses or income have been paid or received.

Accrue. To be obtained, to come from; e.g., profit accrues from a sale.

Accrued expense. Expense, incurred, but not yet paid.

Accrued income. Income, earned, but not yet received.

Acid test ratio (also quick ratio). Quick assets divided by current liabilities; measure of business's ability to pay, expediently, all its current liabilities.

Actual cash value. Actual value of property at time of its loss or damage; frequently, current replacement cost less physical depreciation.

Addenda. Documents, issued prior to opening of bids, clarifying, correcting, or changing bidding documents or contract documents.

Advance billing (also progress billing, overbilling). *See* front-end loading.

Affiliate company (also affiliate). Company that another company controls.

Alter ego. Second or other self; under doctrine of alter ego, court holds individual responsible for acts intentionally done in the name of the corporation; places liability on the individual who uses corporation to conduct his own personal business; liability arises from fraud not against corporation but against people dealing with the corporation.

Amortization. Gradual extinguishment of financial obligation by periodic payments, including an interest charge.

Amortization period (also payback period). Amount of time required to extinguish a financial obligation; number of periodic payments, plus an interest charge, spread over a defined time, necessary to eradicate a financial obligation.

Annual statement. Summary of company's financial operations for a particular year; includes balance sheet supported by detailed exhibits.

Application for payment. Form, with supporting documents that contract documents require, which contractor, requesting payment, issues to owner.

Appraisal. Analysis and evaluation of property to determine its value; an authorized evaluation or estimate of amount of loss.

Appreciation in value. Increase in an asset's fair market value.

Architect. Party qualified, and usually certified as such by statute, to analyze construction projects, create and develop designs compatible with project and properties of materials to be used in project's completion, prepare detailed drawings and specifications, and administer execution of project.

Asset. Entire property of a party subject to payment of party's debts.

Asset current. Cash and resources reasonably expected to be realized, sold, or consumed in normal operation of a business.

Asset fixed. Permanent or semipermanent resources, not available as working capital, needed for or used in successful operation of a business.

Audit. Formal examination and verification of accounts.

Awarding agency. Party for whom work under a contract is done.

Back charges. Billings for costs that a party incurs that, in accordance with agreement, should have been incurred by the party billed.

Backlog. Revenue contractor expects to realize from work to be performed on uncompleted contracts, including new contracts on which work has not begun.

Bad debt. An uncollectable debt.

Bad faith. Design to deceive another; conscious wrongdoing.

Balance sheet. Detailed listing of business's assets and liabilities, representing its financial status at a particular time.

Bankruptcy. Insolvency, under laws of bankruptcy, in which court declares organization subject to having its assets administered to pay its creditors.

Bid bond (also proposal bond). Bond surety issues for contractor competing for a project, guaranteeing to recipient of contractor's bond that if bid is accepted, contractor will execute a contract and provide performance bond; if bid is accepted and contractor fails to execute contract, surety is liable to recipient for amount equal to difference between contractor's bid and bid of next lowest qualified bidders.

Bid date. Date owner establishes on which to receive bids.

Bidding documents. All written materials defining or describing the project or modifying prior written materials for the same project.

Bid security. Financial guarantee to recipient of bid, submitted with a bid, that contractor, if awarded contract, will execute contract in accordance with contract documents.

Bond. Collateral agreement in which one party, called surety, obligates itself to a second party, called obligee, to answer for default of third party, called principal.

Bonding capacity. Total value, based on surety's analysis of total volume of work that contractor can support, of bonds that surety will underwrite for a particular contractor.

Bonding company. Business authorized to issue bonds.

Bond line. Types of bonds that a bonding company issues.

Bond penalty. *See* penal sum.

Book of original entry. Document, recognized by law or custom, in which transactions are recorded and from which postings are made to ledgers.

Broker. Party responsible as general contractor for performance of contract; party enters into subcontracts with others for performance of substantially all construction that contract requires.

Bulldozer. Tractor with a broad, horizontal blade perpendicular to the ground used for pushing excavated materials.

Bylaws. Organization's rules and regulations governing its members and regulating its affairs.

Cash basis of accounting. Method of accounting in which income is recorded only when received in cash and expenditures recorded only when paid.

Cash flow. Origins and uses of a business's cash funds during a particular period.

Change order. Written document, from owner to contractor, issued after effective date of agreement, authorizing contractor to alter the work in a defined manner.

Character. Principal's personal traits, e.g., moral integrity, conscience, surety uses to evaluate risks in prospective bonding.

Charge. An obligation or duty.

Claim. Party's demand for something believed due from another party.

Claims. Amounts exceeding contract price that contractor seeks to collect from parties for their delays, errors, unapproved change orders and other unanticipated losses.

Clarification drawings. Graphics illustrating addendum, modification, change order, or field order; illustrates alterations to original drawings.

Closed shop. A business operation hiring only union employees.

Collateral. Anything of value party pledges to protect in the interests of another party who has obligated itself for debts or actions of first party.

Completed and accepted. Procedure under completed contract method of accounting allowing closing a project on completion of construction and on owner's formal acceptance of project as defined in contract documents.

Completing contractor. Contractor completing project supported by surety because original contractor defaulted.

Comptroller (also controller). Public official or officer of a business auditing accounts and sometimes authorizing disbursements.

Conditions of bid. Terms in invitation to bid, stipulating manner in which bids are to be prepared, submitted, and processed.

Conditions of the contract. Articles in contract defining or describing terms, responsibilities of owner and contractor, performance and payment schedules, and the like.

Conservatism. Understating business's financial affairs to ensure that uncertainties are considered.

Consideration. That which one party does or promises to do; done or promised by one party in return for another party's action or promise.

Construction management. Additional managerial service that architect, engineer, or consultants provide during construction of project.

Construction management contractor. Party to a contract with owner of project; party supervises and coordinates project, including contracting others for construction work.

Contra balances. Balances in accounts that are the opposite of such accounts, e.g., account payable and a debit balance.

Contract. Binding agreement between parties.

Contract administration. Architect's or engineer's managing a project during its construction.

Contract balance. Portion of contract remaining to be completed.

Contract bond. Guarantee that contractor furnishes, indemnifying owner against failure of contractor to comply with terms of contract.

Contract cost breakdown. Contractor's itemized list, prepared after receipt of contract, showing cost of each element and phase of the project.

Contract, cost plus. Provides for reimbursement of defined costs plus a fee, representing profit.

Contract, cost plus, cost plus aware fee. Provides for reimbursement of defined costs plus a two-part fee: (a) invariable amount, fixed at onset of contract, and (b) amount based on performance (e.g., quality, cost-effectiveness).

Contract, cost plus, cost plus fixed fee. Provides for reimbursement plus a fixed fee.

Contract, cost plus, cost plus incentive fee. Provides for reimbursement of costs plus a variable fee, dependent on cost or performance.

Contract, cost plus, cost plus incentive fee, incentive fee based on cost. Provides for reimbursement of costs plus a variable fee, within a defined range, adjusted by a defined formula relating total costs to a target cost established at the outset.

Contract, cost plus, cost plus incentive fee, incentive fee based on performance. Provides reimbursement of costs plus a variable incentive fee based on performance compared to stated performance targets; if targets are surpassed, fee increases; if targets are not met, fee decreases.

Contract, cost plus, cost sharing. Provides for reimbursement of a defined portion of costs, with no additional fee.

Contract, cost plus, cost without fee. Provides for reimbursement of costs with no additional fee.

Contract documents. Agreement, addenda, supporting documents, general conditions, supplementary conditions, specifications and drawings, and modifications to the agreement.

Contract, fixed price (also lump-sum contract). Contract not adjusted for contractor's costs.

Contract, fixed price, firm. Contract not adjusted for contractor's costs or performance.

Contract, fixed price, level-of-effort term. Contract, usually for research and development, obligating contractor to a defined effort for a defined period for a defined, fixed amount.

Contract, fixed price, providing for performance incentives. Provides incentives for contractor to surpass defined performance targets; if targets are surpassed, profit increases; if targets are not met, profit decreases.

Contract, fixed price, with economic price adjustment. Provides for revision of contract price relative to defined contingencies, e.g., fluctuations in material prices.

Contract, fixed price, with firm target cost incentives. Provides at outset a firm target cost, firm target profit, price ceiling, and formula relating final cost to target cost, establishing final price and profit.

Contract, fixed price, with prospective periodic redetermination of price. Provides firm fixed price for initial period with subsequent price redeterminations at defined intervals during remaining period.

Contract, fixed price, with retroactive redetermination of price. Provides ceiling price and, after completion of contract, retroactive price redetermination within ceiling price, relative to performance.

Contract, fixed price, with successive target cost incentives. Provides, at outset, target cost, target profit, price ceiling, formula for establishing the firm target profit, and a time at which formula will be used.

Contract item (also pay item). Defined unit of work for which contract provides price.

Contract limit. Boundary, illustrated on drawings or in other contract documents, on job site beyond which no construction can be done.

Contractor. Party obligating itself to perform a defined project or service.

Contract overrun. Amount by which final contract price, from additional costs from change orders, exceeds original contract price.

Contract payment bond. *See* labor and material payment bond.

Contract performance bond. *See* performance bond.

Contract price. Amount owner pays to contractor, as defined in contract document.

Contract time. Period defined in agreement for completing the work.

Contract, time and materials. Provides payment to contractor based on hours at fixed rates and cost of materials or other defined costs, including a profit factor.

Contract, unit price. Pays contractor defined amount for every defined unit of work contractor performs.

Contractual liability. Another's obligation that a party assumes under contract or agreement.

Corporation. Artificial, legal entity existing separate from its individual stockbrokers; legal "personality" with powers and duties defined in corporation's charter.

Cost of the work. Contractor's cost to perform the work properly.

Cost plus contract. *See* contract, cost plus.

Credit. Ability of business to borrow money.

Current liabilities. Debts or obligations whose payment or liquidation is reasonably expected, requiring expending current assets.

Cycle billing. Dividing accounts receivable ledgers into groups based on common characteristics, e.g., geographical location of customer, and then billing each group at a different time.

Debit. Entry in accounting increasing an asset or decreasing a liability.

Debt–worth ratio. Current debt divided by tangible net worth.

Default. Failure to do that which agreement, duty, or law requires.

Defective work. Work not in compliance with contract documents.

Deferred charges. Expenditures not chargeable to fiscal period in which they were made but carried on asset side of balance sheet.

Deferred income. Income paid in advance but not yet earned.

Depletion. Reduction from removing or using an asset.

Depreciation. Decrease in property's value from use and from obsolescence.

Devaluation of assets. Adjusting downward the value of fixed assets if their current fair market value is significantly below their book value.

Disbursement. Money paid to satisfy a debt or to cover an expense.

Discharge. Removal of obligation or liability.

Disclosure. Act of making known something previously unknown or known to only a few.

Discount. Interest deducted from a note's face value when a loan is made.

Discover. To obtain knowledge for the first time.

Dissolution. Act of dissolving or terminating.

Dissolve. To cancel, terminate, disjoin, abrogate.

Dividend. Portion of corporation's net earnings distributed to stockholders in proportion to number of shares owned.

Division. Part of a corporation created for a defined purpose whose assets are not separate from those of the corporation; corporation is responsible for division's debts.

Double-entry bookkeeping. System recording both debit and credit for each transaction.

Draw. Progress billings currently available to contractor under a contract with fixed payment schedule.

Earned estimates. Figure used in accounting based on estimated amount of work completed within a defined period.

Engineer. Person, firm, or corporation named as such in contract documents; responsible for managing the quality of materials and workmanship and structural integrity of the project.

Equity. Pecuniary value of property exceeding claims and liens against it.

Escalation clause. Contract's article providing adjustments of price of specific items as conditions change.

Estimate (bid function). Costs contractor anticipates for a project; described in contractor's bid proposal.

Estimated cost to complete. Anticipated additional cost of materials, labor, and any other item required to complete a project within a defined period.

Estimator. Party who appraises value, worth, or cost of items, especially construction equipment and materials.

Expenditure. Spending or paying money.

Exposure. Estimate or probability of loss from a hazard, contingency, or condition.

Extras. Work added to original plan, billed separately, and not altering contract amount, that owner requests contractor to do.

Fair market value. Price at which a buyer and seller, under no compulsion to buy or to sell, will trade.

False entry. Entry into accounts, intentionally to misrepresent the truth, to deceive, or to defraud.

Fiduciary. Person occupying position of trust, especially one managing the affairs of another.

Field directive. Written order engineer issues, ordering minor changes in work.

Final acceptance. Customer's certification, from architect or engineer, that project is complete, according to contract documents.

Final inspection. Architect's or engineer's final review of project before owner gives final payment.

Financial analysis. Examination and interpretation of financial statements to evaluate business's financial status.

Financial analysis, comparative analysis. Evaluating trends in successive financial statements.

Financial statement. Written document recording financial operations of a business for a defined period.

Float. (1) Checks credited to depositor's account but not yet debited to drawer's bank account. (2) Time between when a check is written and when it is deducted from drawer's account.

Force account. Work ordered to be done when no agreement for lump-sum or unit price payment is in force; to be paid for flat cost plus overhead and profit.

Front-end loading. Assigning higher values to work to be completed in early contract stages than to work to be completed in later stages to increase cash receipts at outset of project.

General building contractor. Party whose principal business is constructing buildings involving unrelated building trades.

General conditions. Portion of contract documents defining and describing rights and responsibilities of parties to the contract.

General contractor. Party entering into contract with owner and taking full responsibility for project's completion.

General engineering contractor. Party whose principal business is heavy construction, e.g., highways, dams, bridges.

Go-in price. The price at which a contractor begins work on a job, exclusive of any change orders or other price changes that must be made after the work has begun.

Goodwill. In accounting, difference between business's total worth and sum of value of its individual assets.

Guaranteed maximum cost. Amount defined in contract between owner and contractor as the upper limit of money available for completing a project.

Guarantor. Party undertaking that another party will pay or will perform.

Guaranty. Undertaking a collateral contract that another party will pay or will perform.

Incurred but not reported. Liability for future payments on losses occurred but not reported in reinsurer's records.

Incurred expense (other than loss expense). Expense that has occurred but that may or may not have been paid.

Indemnify. To secure against loss, hurt, or damage.

Indemnitor. Party entering into indemnity agreement with a second party, securing second party against loss.

Indemnity. Security against and compensation for hurt, loss, or damage.

Indemnity agreement. Contract entered into between indemnitor and surety in which indemnitor secures surety against loss surety may sustain on bond in behalf of another.

Inflation. Rapid increase in general price level, usually concomitant with increase in supply of money.

Insolvency. Inability to pay debts as due in operation of business; business liability exceeding monetary equivalent of assets.

Inspection. Comparison of actual work with requirements in the contract.

Insurance. Contractual relationship in which one party, the insurer, agrees to reimburse another, the insured, for money or premium or pay for loss on defined subject from defined hazards or perils.

Intangible assets. Valuables such as trademarks and copyrights accruing to a business.

Interest. Money paid for using borrowed money or credit.

Internal control. Organizing a business so that assets, liabilities, expenditures, and revenues can be properly managed.

Investment income. Money earned from invested assets, including realized capital gains.

Invited bidders. The only contractors from whom owner will accept bids.

Joint venture. Contractors combining their skills and financial resources to undertake construction contracts for mutual benefit of participants.

Kiting. Writing check against bank account having insufficient money to cover check, hoping that money will become available before check is deducted from drawer's account.

Labor and material payment bond (also payment bond, labor and material bond). Bond that contractor gives in which surety guarantees to owner that contractor will pay for labor and materials used in executing the contract.

Lease-hold. Right to use property by power of lease.

Lessor. Party transferring property by lease.

Letter of agreement. Letter from one party, the addresser, to a second party, the addressee, stating terms of agreement between the two parties; addressee's signing is acceptance of defined terms as legally binding.

Letter of intent. Letter stating intention to enter into formal agreement; usually stipulating agreement's requirement.

Liability. Legally enforceable obligation.

Lien. Charge court issues on party's property to discharge a debt or duty.

Liquid. Capable of being readily converted to cash.

Liquidated. Ascertained, determined.

Liquidation. (1) Conversion of assets to cash. (2) Determining by agreement or litigation exact amount of indebtedness.

Liquidation value. Value of property sold to settle a debt.

Losses outstanding. Losses occurred but not paid.

Lowest responsible bidder. Contractor who submits lowest bid and, additionally, whom owner considers to be least risk.

Lump sum. Money given in one payment.

Main contractor. *See* prime contractor.

Market price. *See* fair market value.

Market value. *See* fair market value.

Monitoring. Administering activities of personnel, financial handling of accounts, management of projects, and the like.

Negligence. Failure to exercise level of care that reasonable and prudent person would be expected to exercise in same circumstance.

Negotiated contract. Agreement for construction developed without competitive bidding by negotiating specifications and terms.

Net operating income. Income before interest and income taxes but after depreciation.

Net quick (also net quick assets, working capital, net working capital). Current assets exceeding current liabilities.

Obligee. Party that bond secures against loss.

Obligor (also principal). Party liable under contract of suretyship.

Observation of the work. Architect's function during project's construction; to determine if work is being performed according to contract documents.

Obsolescence. Devaluation of product's functional or physical assets or value, from technological changes, not from deterioration.

Operating cycle. Circulation of current assets in normal business.

Opinion, auditor's disclaimer of. Auditor's opinion, after analyzing business's financial records, that he is unable to express an independent opinion because he was unable to use all normal tests of accounting records.

Opinion, auditor's qualified. Auditor's opinion, after analyzing business's incomplete financial records, that records examined fairly reflect business's financial status for period examined.

Opinion, auditor's unqualified. Auditor's opinion, after analyzing business's financial records, that records examined accurately reflect business's financial status for the period examined.

Overbilling. *See* front-end loading.

Overhead. Business expenses, e.g., rent, insurance, heating, required for continuing operation of a business.

Owner. Owner of project under construction for whom contractor performs construction work.

Ownership equity. Residual interest in assets less liabilities.

Partnership, general. Joint venture in which individuals, each of whom is responsible for business acts of the others, combine resources to engage in business, usually permanently.

Partnership, limited. Joint venture, organized under Uniform-Limited Partnership Act, in which limited partners, each with cash or tangible property investment, have a limited liability in business operations.

Payables. That to be paid.

Penalty clause. Article in construction contract providing for reducing amount defined in contract to contractor as penalty for failing to meet defined targets or for failure to meet contract's specifications.

Penal sum (also bond penalty). Maximum amount for which surety is held liable under bond.

Percentage-of-completion method. An accounting system used on certain long-term contracts, in which profit is accrued over life of contract according to progress made.

Performance bond. Guarantee faithful performance of terms of written contract; secures owner against contractor's failure to perform contract.

Plans. Graphics that owner approves, showing details of work to be performed.

Prebid conference. Meeting held between owner and prospective contractors before bidding, allowing contractors to familiarize themselves with the project and to learn any of owner's further requirements.

Prepaid expenses. Cost incurred, frequently recurrently, to obtain goods or services not yet received, consumed, or expired.

Prequalifications. Owner's written approval authorizing contractor to bid on a project when bidders are required to meet certain defined standards.

Prime contract. Agreement between owner and contractor for completion of a project.

Prime contractor. Contractor entering into contract with owner of project for completion of that project.

Principal. Party primarily liable under a bond of suretyship.

Procurement. Act of obtaining or securing something, e.g., supplies, work.

Profit and loss statement. Financial statement showing contractor's profits and losses for a defined period.

Pro forma statement. Financial statement for an anticipated, but not yet completed, transaction.

Progress billing. Amounts contractor bills owner, according to contract, based on progress to date on the project.

Progress chart. In critical path method of planning projects, graphics, illustrating operations needed to complete a project; includes proposed starting and completion dates.

Progress payment. Money owner gives to contractor, according to contract, based on progress to date on the project.

Project manager. Individual or consulting firm surety hires to administer completion of construction contracts and to document activities on the project for surety.

Punch list. List made near completion of project indicating items contractor or subcontractor must do to complete work in accordance with contract.

Purchase order. Written document authorizing purchase of materials or services.

Quantity takeoff. Itemized list of materials and labor contractor requireq for completing a project; contractor uses list to prepare a bid.

Quick assets. Liquid assets, e.g., cash, accountsreceivable, marketable securities, that can be easily and rapidly converted to cash.

Quick ratio. *See* acid test ratio.

Quotation (also quote). Price, submitted prior to opening bids, for which subcontractor or supplier agrees to supply services or materials to a contractor should contractor be awarded contract.

Quote. *See* quotation.

Receivable. That which is owed to a party.

Replacement cost. Cost to substitute equivalent items for equipment or services.

Requisition. Written request from one department of a company to another department, e.g., purchasing, central supplies, for a defined item.

Retainage. Contract balances held back for payment to contractor on completion of contract.

Retained earnings. Corporation's net profits not paid out as dividends.

Retentions. Amount owner withholds from progress billings until contractor satisfactorily completes project.

Return on investments. Money earned from money invested.

Reversing entries. In accounting, entries into a journal at the beginning of a fiscal period to reverse entries made at the end of the prior fiscal period.

Salvage value. Asset's value after asset's useful life has expired.

Schedule. A list.

Schedule of values. Document showing the amount of the total contract sum to be allocated to each phase; owner gives schedule to architect or engineer who uses it to evaluate contractor's application for payments.

Shop drawings. Graphics contractor, subcontractor, manufacturer, or supplier prepares illustrating material or equipment for work or for work itself.

Solvency. Ability to satisfy all financial obligations.

Specialty contractor. One whose business is a specialized trade, e.g., heating and cooling, plumbing.

Specifications. Complete, detailed written description of all materials and work required to complete a project.

Subcontract. Contract between party to an original contract and another party, especially contract to provide work provided in original contract.

Subcontractor. Party contracting to perform part or all of another's contract.

Substantial completion. Point at which engineer expresses in his certificate of substantial completion that work is sufficiently complete, as defined in contract documents, to be utilized for its intended purposes.

Supplementary conditions. Part of contract documents adding to or modifying the general conditions.

Supplier. Party providing or furnishing commodities.

Surety. Party collaterally liable for payment of money or performance by another party.

Take off. Estimating amount of material needed for a particular project.

Unearned income. Income received but not yet earned.

Vendor. Person selling any property, especially real estate.

Work. Entire construction or parts of construction contractor must perform in accordance with contract documents.

Working drawings. Graphics illustrating in sufficient detail work to be done that the work can be performed with no other instructions.

Work on hand. Amount of money contractor has remaining in his contracts to complete all projects; total estimated cost to complete, excluding advance payments to be earned and amount of losses.

About the Author

Thomas C. Schleifer has worked in construction since age 16. He brings the lessons and experience of over 30 years in contracting and consulting to the preparation of *Construction Contractors' Survival Guide*.

He attended Newark College of Engineering and Fairleigh Dickinson University Graduate School of Business.

His experience includes serving as foreman, field superintendent, then vice president and co-owner of a construction company.

From 1976 to 1986 he was President, then Chairman of the largest international consultancy firm serving the contract surety industry. During this period he assisted in the resolution of hundreds of troubled or failed construction firms.

This combination of practical, hands-on experience as a contractor and assisting troubled construction companies has given him a unique perspective on the reasons why some contractors prosper while others do not and the ways to minimize and avoid the risks that can lead to failure.

Currently, Tom spends his time advising contractors on organization, structure, and strategic planning while he writes, lectures, and teaches.

The importance of education in the construction industry is one of Mr. Schleifer's favorite themes. He is a former Chairman of the Construction Education Subcommittee on Continuing Education of a national association. He has lectured extensively at universities and professional and trade associations and has authored numerous articles. He is active in a number of professional and trade associations and is listed in "Who's Who in Finance and Industry" and "Who's Who in America."

Other publications by Tom Schleifer include "Glossary of Suretyship and Related Disciplines" and "The Prudent Underwriter" (videotape).

INDEX

Printed and bound by CPI Group (UK) Ltd, Croydon, CR0 4YY

23/04/2025

14660922-0005